ALARMS

55 Electronic Projects and Circuits

ALARMS

55 Electronic Projects and Circuits

Charles D. Rakes

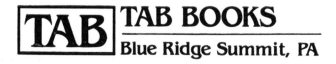

TAB BOOKS
Blue Ridge Summit, PA

FIRST EDITION
FIFTH PRINTING

© 1988 by **TAB Books.**
TAB Books is a division of McGraw-Hill, Inc.

Library of Congress Cataloging-in-Publication Data

Rakes, Charles D.
   Alarms : 55 electronic projects and circuits / by Charles D.
Rakes.
      p.   cm.
   Includes index.
   ISBN 0-8306-2096-6   ISBN 0-8306-2996-3 (pbk.)
   1. Electronic alarm systems—Design and construction.   I. Title.
TK7241.R34   1988
621.389'2—dc19   88-16020
                                            CIP

TAB Books offers software for sale. For information and a catalog, please contact
TAB Software Department, Blue Ridge Summit, PA 17294-0850.

CONTENTS

INTRODUCTION

There's no turning back the clock when it comes to the stampeding advances made daily in electronics. For any hope at all of keeping abreast of at least the major accomplishments, you must use your allotted time wisely or be lost forever on a slippery treadmill, falling farther and farther behind. One of the best ways to learn about any subject is to have hands-on experience along with the book studies. The ideal combination is one that offers practical circuits and projects that are not only fun and educational to build, but fulfill a specific need.

The majority of us move and work in a world that is operated and controlled by electronics. The automobile is used daily, and without the complex electronic support the car would be no more than a bucket of bolts rusting away in some obscure junkyard, leaving us all pedaling around town. Without electronics, our home life would be taken back to the times of our great-grandparents, and the modern convenience missed would fill a book. We live in an electronically supported world. Those of us who work with electronics as a hobby or profession can use our skills to keep our own equipment going.

The circuits and projects in this book are designed to help keep minor problems from turning into major breakdowns. Almost everything that functions can be monitored by an electronic circuit to establish its operational condition, and to report with an alarm output when a set limit is breached. All of the circuits and projects in this book have that goal: to monitor and evaluate the data, and to send out an alarm in time to take corrective measures and avoid a major problem.

Each of the circuits and projects presented in this book has been built, tested, and carefully debugged to perform the job it was designed to do, so dig in and enjoy building your own electronic projects.

Chapter 1
Alarms, Sensors, Sounders, and Control Systems

Today's high-tech world is filled with electronic and mechanical devices that help to make our lives easier. But let one simple circuit component fail in any of the complex electromechanical systems that surrounds us and our modern lifestyle can come to a screeching halt.

AUTO ALARMS

A modified system of checks and balances is often used to avoid a complete shutdown of a critical piece of equipment, such as the family automobile. Most of the major critical, functioning mechanical and electrical components are monitored by sensors that feed status indicators. If you keep a watchful eye on the various status indicators in an automobile (including gauges, meters, lights, and audio sounders) you usually can avoid a major breakdown and perform proper preventive maintenance.

Unfortunately, not all vehicles come factory-equipped with all of the desired sensors and indicators. With a little time, a few dollars, and the desire to upgrade your own tin horse, however, the circuits and projects in Chapter 2 should fill the bill.

Let's take a brief look at the types of sensors and indicators that can be homebuilt and used in the family car. One of the most important parts used in today's automobile is the wet cell battery. Not only does this battery start the car's engine, but it also supplies electrical power to the ignition system, onboard computer, electronically controlled carburetor or fuel injectors, lights, entertainment equipment, and power for all sensors and indicators.

1

As important as the car's battery is to the total operation of the vehicle, you are truly lucky if the manufacturer supplies a simple idiot light to indicate that the charging system is doing something. A very few cars come equipped with an analog expanded scale volt meter that gives you a better indication of the condition of the car's battery. To take advantage of this feature, however, you must know just how to interpret the meter's reading. Because not everyone knows how to interpret the meter's action, you can build a simple high/low battery alarm circuit to do the job for you. (See Chapter 2.)

If an idiot light comes on and is noticed in time, a major breakdown can usually be avoided, but what happens if the idiot light bulb is kaput? Build the lamp burn-out alarm in Chapter 2 and avoid any chance of this problem happening to you.

You can add many additional sensors and alarm systems to today's modern tin horse to insure a longer and more useful life for you and your car. They include a cooling system over heat alarm, a back-up alarm, an intrusion alarm, and several other specialized alarm systems.

INTRUSION ALARMS

One of man's earliest safeguards against life-threatening conditions was the ability to sense the presence of oncoming danger before it happened, and make the proper correction. This vanguard of defense was accomplished with one or more of the five basic human senses: sight, hearing, smell, taste, and touch. Even in today's advanced technical world with all of its sophisticated electronic equipment, the final decision as to what action to take is based on what your senses tell you. Chapter 3, *Intrusion Alarm Sensors*, offers circuits and projects that use sight and sound to help avoid a break-in or burglary.

The most common sensor used in intrusion alarm systems is the ultra-simple make-and-break circuit sensor. This type of sensor is primarily used to protect doors, windows, crawl spaces, safes, and file cabinets, as well as in any application where an electrical circuit is opened or closed by a mechanical device. This type of sensor includes trip wires, magnetic switches, mechanical switches, contact switches, metal foil tape, and other specialized make/break circuits. Figure 1-1 shows a number of make/break sensors connected to a control unit.

The basic light alarm sensor offers a number of interesting and useful circuits that can be built and used with one of the control systems in Chapter 4. Figure 1-2 illustrates the use of a simple single-light sensor. A light source is located at one end of the protected area and is aimed toward the photocell located at the opposite end of the protected area. As long as the light hitting the photocell is not interrupted, the sensor remains in its normal no output condition. If, however, an object passes between the light source and the photocell the sensor's output switches to an alarm condition.

An unusual expansion of the single photocell sensor to a multicell pickup is illustrated in Fig. 1-3. A single nondirectional light source is positioned in the center of the area to be protected. Surrounding the light source are a number of directional photocell sensors that are aimed toward the single light source, providing a near perfect blanket of protection (depending on the actual number

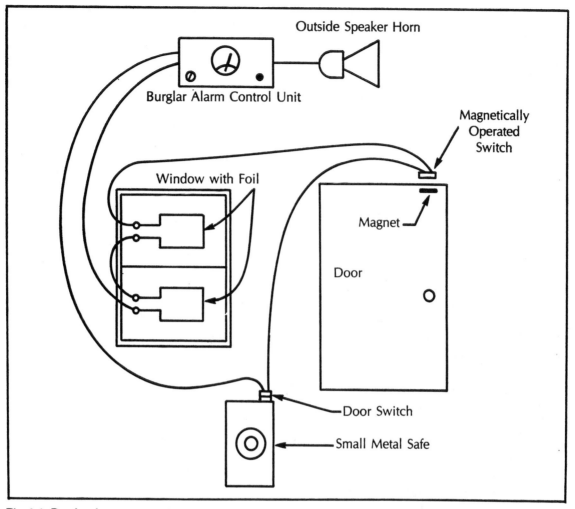

Fig. 1-1. Burglar alarm system using the make/break sensors, with several break-type sensors in use.

of sensors used) for the entire area. The light sensor is an excellent pickup for an advanced alarm system.

The proximity detector circuit is a specialized type of sensor that can be used to protect file cabinets, safes, metal doors, and most other metal items. Figure 1-4 shows a typical application of the proximity sensor protecting a large metal safe.

Figure 1-5 shows one type of proximity detector circuit. The output of a low-frequency oscillator (10 kHz to 100 kHz) is coupled through two variable capacitors, with the center connection going to the protected object or a sensor pickup. A detector circuit is connected to the output of the oscillator through both of the variable capacitors. The two capacitors are adjusted to a point where the detector circuit receives just enough energy from the oscillator to keep the relay energized. If an object is brought into close proximity of the pickup sen-

3

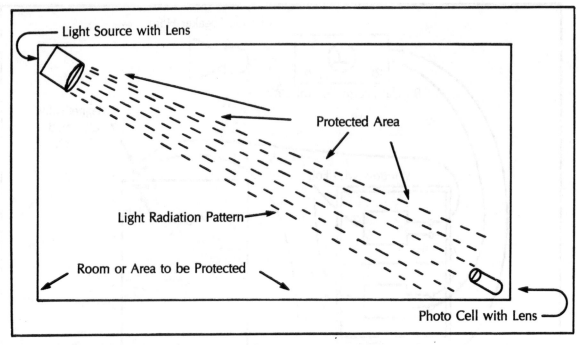

Fig. 1-2. Light source and photo cell burglar alarm sensor.

sor plate, a portion of the rf energy is capacitor-coupled to that object, lowering the amount that is available to the detector circuit. The detector circuit is now receiving too little rf energy to keep the relay closed, causing it to drop out and giving an alarm output.

A properly designed proximity sensor circuit of this type can pick up an intruder's movement several inches from the protected item. The intruder cannot use thick gloves to foil the alarm system.

Another specialized sensor is the ultrasonic motion detector that can be used to protect an entire room or hallway from an outside intruder. Figure 1-6 shows an ultrasonic sensor that uses the Doppler effect to detect movement within its radiated field. A block diagram of an ultrasonic sensor using the Doppler effect is shown in Fig. 1-7.

An ultrasonic oscillator (15 kHz to 45 kHz) drives a specially designed transducer to make up the transmitter (tx) section of the sensor. The transmitter's field is radiated in a pattern similar to the one shown in Fig. 1-6. A portion of the signal is reflected back to the receiver's pick-up transducer and amplified to a level sufficient to drive the mixer circuit. A reference signal is taken from the tx circuit and is fed to the other input of the mixer. If the tx's radiated field is without interference from a moving object, the signal feeding both inputs of the mixer is the same frequency. Under this normal circuit condition, the output of the mixer is zero. If an object is moved within the tx's field, the reflected signal seen by the receiver is shifted in frequency by the amount determined

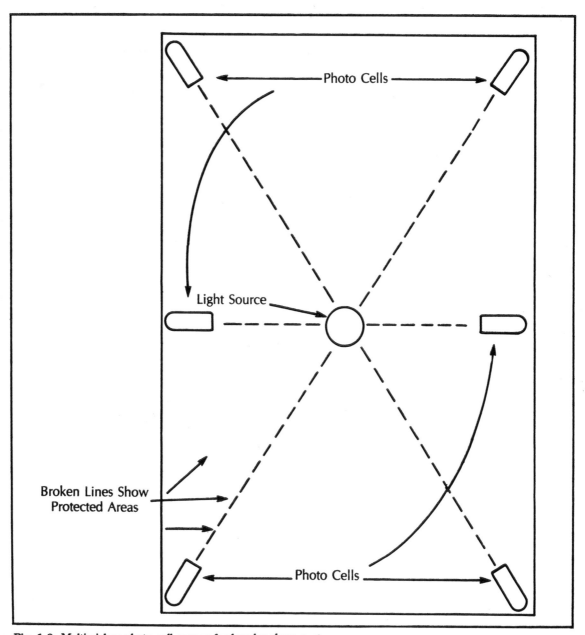

Fig. 1-3. Multi-pickup photo cell sensor for burglar alarm system.

by the speed of the moving object. This frequency shift is known as the *Doppler effect*.

The classic example that best describes the Doppler effect is the story about a man, many years ago, who sat at a train station waiting for a southbound train. While sitting outside the station on that warm summer day, he noticed that, as the nonstop trains steamed by, the trains' whistles seemed to increase in

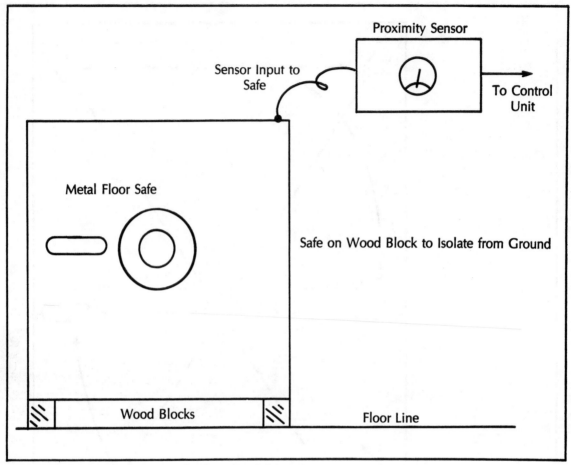

Fig. 1-4. Proximity sensor connected to large metal floor safe.

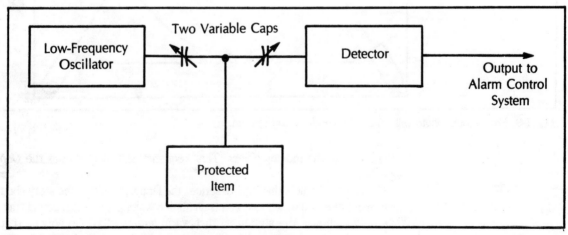

Fig. 1-5. Block diagram of a proximity alarm sensor circuit.

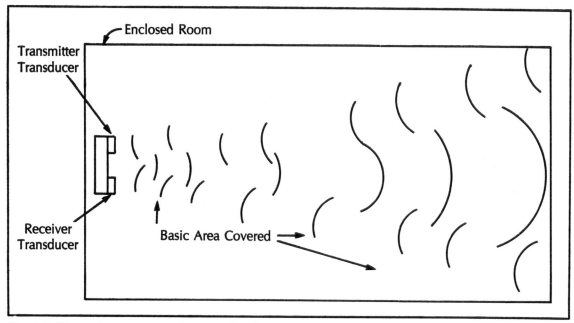

Fig. 1-6. Ultrasonic sensor protecting an enclosed area.

pitch as they approached and decrease in frequency as they sped by. You can experience the same Doppler effect, to a somewhat lesser degree, if you listen to the siren of an emergency vehicle as it approaches and passes by. The ultrasonic sensor is just one of many uses of this unusual sound effect.

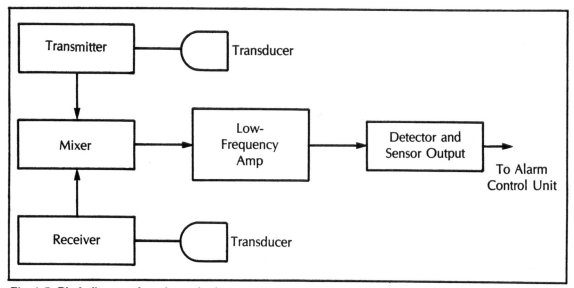

Fig. 1-7. Block diagram of an ultrasonic alarm sensor.

Several other specialized intrusion alarm sensors are covered in Chapter 3, including sensors designed to pick up vibration, or to pick up the passing of a motor vehicle; sound-activated sensors; and other sensors.

BURGLAR ALARMS

Chapter 4 offers a number of burglar alarm control systems that are designed to operate with the sensors in Chapter 3 and the alarm sounders in Chapter 5.

The first burglar alarm control system is designed to be used with the basic make/break input circuits. See Fig. 1-8 for the block diagram of a single-input control system. The control unit shown in Fig. 1-8 is built for the normal open-circuit sensors. In this type of input circuit, all of the sensors are wired together in parallel. When any one or more of the sensors closes, the alarm is set off.

The block diagram in Fig. 1-9 illustrates the control unit designed to operate with the normally closed sensor circuit, or *break circuit*, and are all wired in series. If any one of the input sensors open, the alarm is set off.

A block diagram of a multi-input alarm control system is shown in Fig. 1-10. This control unit responds to both types of sensors, and if any of the sensors change their normal state the alarm is set off. A number of other types of alarm control units are discussed in Chapter 4. They have timed entry/exit, multi-input/output, and other useful features.

Chapter 5 discusses alarm sounders and indicators. Circuits for low-level sounders for indoor use and extra high level output sounders for outdoor applications are covered, with several variations to fill almost any alarm requirement.

The vast telephone system network throughout the world can be used to transfer alarm status information from one location to another, giving an almost

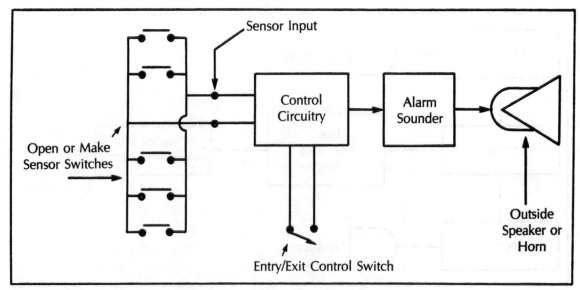

Fig. 1-8. Normally open sensor burglar alarm control system.

8

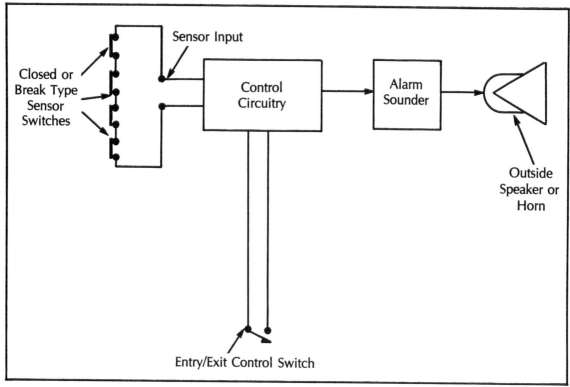

Fig. 1-9. Normally closed sensor burglar alarm control system.

unlimited access to interrogate an alarm system from anywhere on earth. Chapter 6 offers a telephone remote-monitoring control system that takes advantage of the large telephone network.

VOLTAGE ALARMS

A number of unusual alarm circuits and projects are offered in Chapter 7, including an over/under voltage alarm circuit that can be used to monitor a battery's output or an ac-operated dc supply.

For a number of electronic devices, an under-voltage circuit condition can be more of a problem than a slight over-voltage condition. A computer can hum right along looking good, but let the ac input or dc output sag for a short period of time and zap goes a load of data. If the low-voltage problem is a result of a slowly failing part in the power supply, one of the voltage sensor alarms can give ample warning before the zap occurs. This early warning could save a day's work, and also let you make the repair before the problem escalates into something more serious.

Another circuit condition can cause as much, or even more, damage than either of the previous two (in some cases it is caused by an over/under voltage condition): a circuit that slowly consumes more and more current until something smokes. Chapter 7 offers a solution to the hungry circuit that uses too

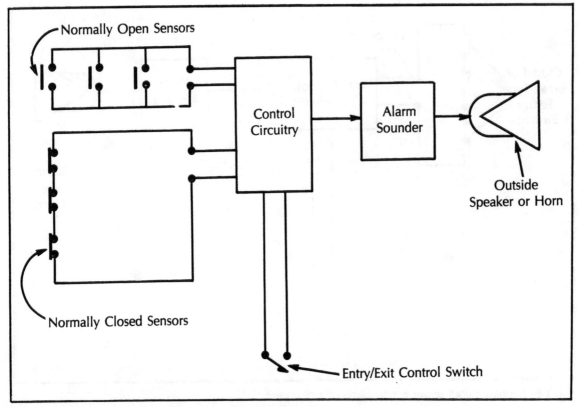

Fig. 1-10. Multi-input sensor burglar alarm control system.

much current. This control tattles in time to save both the power source and the equipment that it's feeding.

One of the over-current circuits takes the warning one step further by automatically shutting down the supply until the problem can be corrected. This feature can save the equipment without the need for someone to monitor the operation.

OTHER ALARMS

Another useful alarm circuit is the temperature sensor which sounds off if the heat goes beyond a preset limit. This type of sensor and alarm also can be used to protect expensive electronic or mechanical equipment that can be damaged by an overheated condition. Many times the overheating in electronic equipment can be traced back to a power supply failure, so the temperature alarm and over/under voltage alarm systems can work hand in hand to double-check the condition of an expensive piece of equipment.

Circuit sensors that are turned on by moisture are useful in protecting all types of equipment from water damage. They can also automatically start a water pump to clear an area of moisture. A leaky basement might be a good place to use the combination.

Tornadoes and dangerous thunderstorms have been known to strike every one of the 48 continental states at one time or another. There is little, if anything, that can be done about the weather, but you can be forewarned about many of the oncoming storms and take appropriate action to save life and limb. A severe-weather warning alarm is covered in Chapter 7. It can help determine the direction and intensity of an oncoming storm.

Chapter 2
Automobile Alarms

The double-digit inflation of the late 1970s has pushed the cost of the family automobile up into second place in the average family budget, following only the expense of the home. It doesn't take a genius to realize that a dose of preventive maintenance is in order just to hold down the expanding cost of owning a vehicle, and that is exactly what each of the following automobile alarm circuits are designed to do.

A BATTERY ALARM

If any single item in an automobile could be referred to as its heart, it is the car's battery. Almost every important functioning part of the modern automobile requires a dc source to operate, and when the battery goes the car stops.

There's no magic formula or secret method that can keep a car's battery going indefinitely, but a battery that's going bad because of poor service can be monitored by an electronic sensor circuit to raise an alarm when a problem is just beginning to develop. Of course, if the battery is several years older than its manufacturer's warranty date, the alarm could be signaling that the battery is dying a natural death. On the other hand, it could be a warning that the battery's liquid level is low, or there is a current overload, a dirty cable terminal, a loose alternator belt, a failing alternator, or some other battery-related fault— any of which could cause the car to stop or fail to start. Most of these problems, caught in an early stage, can be corrected before an unplanned stall occurs resulting in a towing charge.

An excellent way to keep tabs on the battery's condition is to constantly monitor its voltage under all load and charging use conditions. The circuit shown in Fig. 2-1 does just that by monitoring a preset low-voltage level and a preset high-voltage level.

If the battery voltage goes above or below the preset limits, an audible alarm sounds a warning of an unusual battery condition. An LED is used for each of the limits; the one indicating the out-of-limit voltage level turns on, showing how to locate and correct the problem.

How the Circuit Works

The monitoring circuit and the sensor inputs are the same; both connect across the car's battery (see Fig. 2-1). This not only supplies power for the circuit operation, but also allows for a direct input to both of the voltage monitoring circuits.

Zener diode D1 supplies a 5.6-volt dc regulated source that both the low- and high-limit circuits use as a reference voltage. Transistor Q2's base is connected to a wiper of a 10-turn trim pot that is connected across the car's battery. Q2's emitter is tied to the reference zener D1. The collector of Q2 is direct-coupled to the base of Q1 through a current-limiting resistor, R2. Q1's collector drives an LED indicator through a current limiting resistor, R3, and supplies voltage to the audible sounder through diode D4.

The emitter of Q2 always remains at 5.6 volts, and the base voltage varies with any change in the battery voltage. If the trim pot, R5, is adjusted to a point just below the bias level needed to turn Q2 on while the battery is at its maximum safe voltage level, then the circuit is properly set for the HIGH alarm battery condition. A small increase in the battery's voltage over the preset level causes Q2 and Q1 to conduct, lighting the LED and sounding the alarm to indicate an over-voltage battery condition.

The low-voltage alarm circuit is similar to the over-voltage circuitry, but functions in a reverse manner. The zener D1 also supplies the same 5.6-volt reference voltage to the low-voltage alarm circuit at the emitter of Q5. Under normal battery voltage conditions, with R6 preset for a minimum voltage limit, Q5 and Q4 are biased on, holding Q3 in the off condition. If the battery voltage drops below the preset setting, Q5 and Q4 turn off, allowing Q3 to draw base current through R7, thus turning on the LED and sounding the alarm.

Setting up the circuit limits is best done on the bench with an accurate voltmeter and a variable power supply. More about this follows the construction information.

Building the Circuit

You can construct the circuit on a piece of perf board, etched circuit board, or whatever works best for you because the circuit is noncritical, and the parts

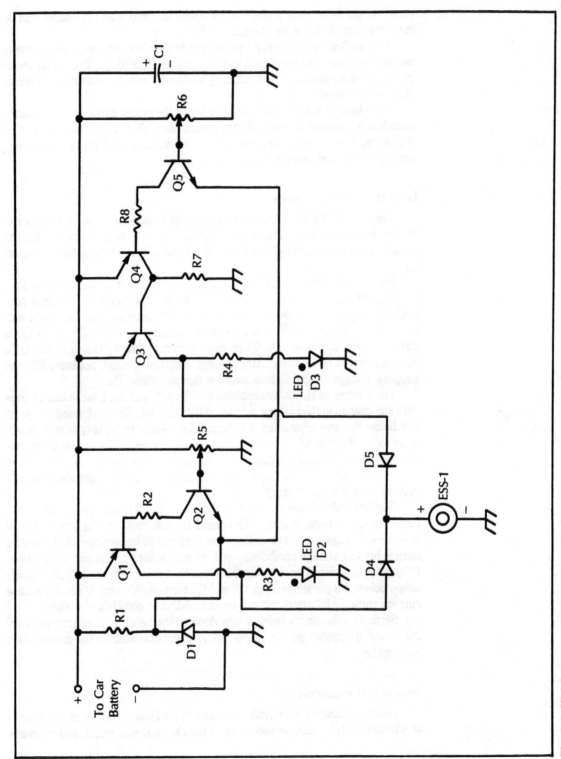

Fig. 2-1. High/low battery voltage alarm circuit.

Table 2-1. Parts List for Fig. 2-1.

C1	100 mF 25V electrolytic capacitor
D1	1N4734 5.6V ½ W zener diode
D2, D3	Red LED diode
D4, D5	1N918 signal diode
ESS-1	Solid-state electronic sounder
R1, R3, R4	1 kΩ ½ W resistor
R2, R8	2.2 kΩ ½ W resistor
R5, R6	10 kΩ mini 10-turn trim pots
R7	4.7 kΩ ½ W resistor
Q1, Q3, Q4	2N4249 or 2N3638 pnp silicon transistor
Q2, Q5	2N2222 npn silicon transistor
Misc.	Perf board, terminals, cabinet, wire, etc.

layout requires no special attention. Of course, you must make sure all connections are electrically and mechanically sound, and a degree of neatness always helps in making an adjustment or in troubleshooting a circuit. You can house the circuit in any suitable metal or plastic enclosure. If you want the completed unit to be mounted in full view, with the two LEDs facing the driver, use a clean and neat enclosure that blends in with the other car instruments. If, however, you want only the LEDs to be mounted in a visible location, you can use any type of enclosure to house the circuit.

Figure 2-2 shows a suggested layout using a 2½-×-4-inch section of perf board to mount the component parts. You can mount the two LEDs and the audio sounder to the front part of the housing cabinet or mount them remotely elsewhere, whatever works out best for the particular installation.

Setting the Two Voltage Limits

You will need a digital or an accurate analog dc voltmeter and a variable dc power supply to do a first-class job of setting the alarm voltage limits. First, set both of the trim pots to their mid positions and connect the power supply to the alarm's circuit battery input. Connect the voltmeter across the power supply's output and adjust for 15 volts. The actual upper voltage limit can vary somewhat from the suggested 15-volt setting, but if you set the limit too high, it could cause a missed warning of a problem with the battery's charging system. Therefore, take care in selecting the exact voltage value used in the high-level adjustment.

After you have selected the desired high-voltage limit level and connected it to the circuit's input, adjust the trim pot R5 until the LED (D2) just lights and the audio sounder begins to sound off. To check the exact high-voltage setting, lower the supply's voltage to 10 or 11 volts, and slowly increase the supply's voltage, noting the meter reading when the LED lights. After a couple of slight pot adjustments, you can set the exact voltage limit.

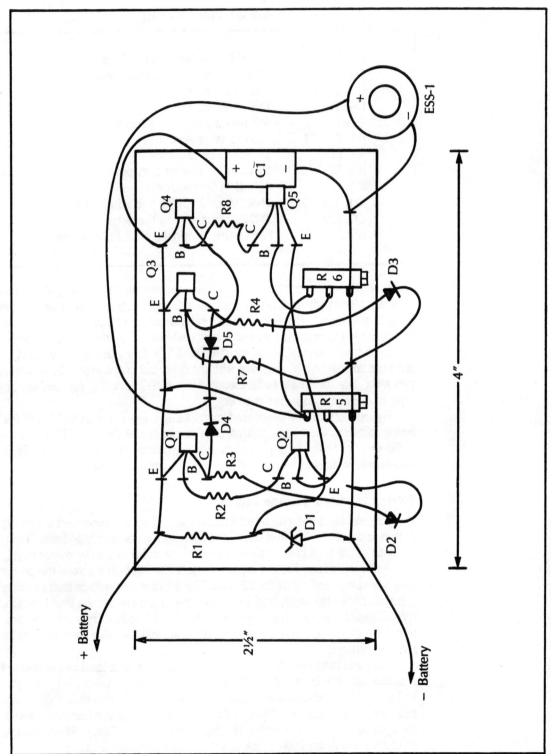

Fig. 2-2. Component parts layout for H/L voltage alarm.

If the low-battery alarm voltage is set too high, the LED and sounder turn on when the starter is turning over the car's engine. Actually this is not a bad feature because it lets you know that the low voltage circuit is working and doing its job.

Set the power supply to 11 volts, or whatever voltage level you want for the low limit, and adjust the trim pot R6 until the LED (D3) just begins to glow. Run the supply voltage back up to about 12 volts and then slowly lower the supply voltage, while keeping an eye on the voltmeter, until the LED lights. This completes the circuit calibration.

Wire the monitoring circuit into the car's electrical system so that when the key is turned off the circuit is turned off also. There is no need to draw current from the battery when the car is not in use.

A LAMP BURN-OUT ALARM

A car's taillight can be out for some time before it's discovered by the driver or police officer. One method of discovery can cost the price of a replacement bulb, but if an officer relays the information it can cost much more than the price of the lamp.

A simple lamp burn-out alarm circuit is shown in Fig. 2-3. A cadmium sulfide photocell is located close to the lamp being monitored. While the lamp is on, the internal resistance of the P.E. cell is very low. With the P.E. cell connected to the base of Q1 and the negative side of the circuit, the bias current flowing through R1 is passed to ground through the low resistance of the P.E. cell. The transistor is biased off and no current can flow through the solid-state sounder. If the lamp burns out or fails to light, the P.E. cell's resistance goes

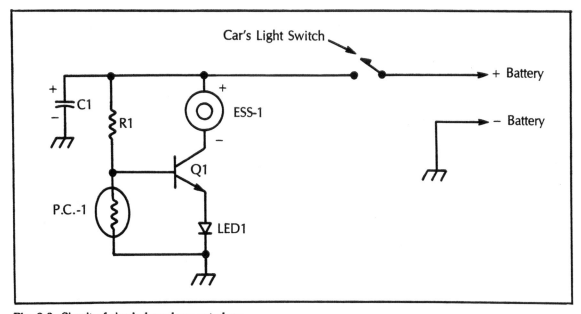

Fig. 2-3. Circuit of simple lamp burn-out alarm.

Table 2-2. Parts List for Fig. 2-3.

C1	100 mF 25V electrolytic capacitor
LED-1	Red LED
R1	47 kΩ ¼ W resistor
PC-1	Cadmium sulfide photocell
Q1	2N2222 npn silicon general purpose transistor
ESS-1	Solid-state electronic sounder
Misc.	Perfboard cabinet, wire, solder, etc.

way up in value, allowing the current to flow into the base of Q1, turning it on, lighting the LED, and activating the solid-state sounder. The power source for the circuit is taken from the same connection where the power is switched on and off for the lamp. Taking the power from this point keeps the circuit from sounding off when the lamp is turned off.

Building and Using the Simple Burn-Out Circuit

You can construct one or more of the single-sensor input burn-out circuits on perf board and house it in a small plastic cabinet. Place the LED indicator and sounder in a convenient location that the driver can easily see and hear without interfering with the safe operation of the vehicle. Use whatever construction scheme best suits the application. Locate the P.E. cell as close to the light source as possible, facing the light source.

An expanded multi-input lamp burn-out alarm circuit is shown in Fig. 2-4. With this circuit, a total of six individual lamps can be monitored at the same time, all connected to a common alarm sounder. If any one of the lamps fails, an LED lights, indicating which lamp failed, and sounds the audible alarm. Normally the number of lamps that is always on at the same time, in most applications, does not add up to six.

You can use less than six sensors either by leaving off the input and output component parts to the unused inverters, or if a future need is possible, by just connecting a temporary jumper across where the P.E. cell would wire, and leaving it in place until the circuit sensor is needed.

If a sensor stage will never be placed in service, leave off the pc cell, output LED, 1N914 diode, and 1 k output resistor, but leave on the input 27 k resistor that is connected between the inverter's input and circuit ground. This grounded input keeps the unused inverter stages from any unnecessary activity and possible internal damage.

Before you make any additional circuit changes, too, look at how the multi-input alarm circuit operates. This check may help you build and use the alarm. Like ducks lined up all in a row, the six sensor circuits are alike, with separate input and output circuits. All of the sensors' output circuits are diode-coupled to a common bus to supply power to the solid-state sounder. Because all sensor circuits are alike, a description of sensor circuit A serves for all six. A light-

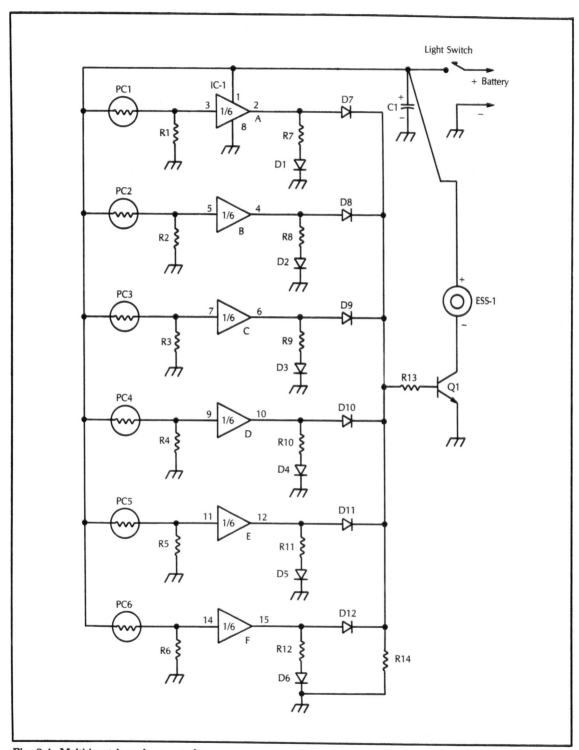

Fig. 2-4. Multi-input lamp burn-out alarm.

Table 2-3. Parts List for Fig. 2-4.

C1	100 mF 25V electrolytic capacitor
D1-D6	Red LEDs
D7-D12	1N914 silicon signal diodes
ESS-1	Solid-state electronic sounder
IC-1	4049 inverting hex buffer
Q1	2N2222 npn silicon general purpose transistor
R1-R6	27 kΩ ¼ W resistor
R7-R12	1 kΩ ½ W resistor
R13	10 kΩ ¼ W resistor
R14	100 kΩ ¼ W resistor
PC1-6	Cadmium sulfide photocell
Misc.	Cabinet, perf board material, push-in pins, wire, solder, etc.

activated P.E. cell is connected to the input stage of an inverting hex buffer IC, causing the input to be in the high state. The output of an inverter is always opposite to its input state, so the output is now in a low state, or near zero voltage.

As long as the voltage at the output of the inverter remains low, the LED (D1) will not light and there will be no forward bias to turn Q1 on and sound the alarm. If the lamp supplying light to the P.E. cell fails to light, the inverter will see a low-voltage state at its input. As a result, the output will go high, lighting the LED (D1) and supplying bias through D7 to the base of Q1, and turning on the solid-state sounder. The LED and solid-state sounder will remain activated as long as one or more of the inverter stages output is high and power remains on to the alarm circuit.

Building the Multi-Input Alarm

Here again the circuit is not in the least critical, so any good construction scheme will do. Perf board and push-in pins are a good choice for mounting the circuit components on, but you could make a p.c. board for the circuit—just use whatever construction method you feel the most comfortable with. Take extra care in locating and mounting the P.E. cells next to the lamp under check. You can use a little dab of silicone rubber to keep the P.E. cell in place without damaging either the cell or mounting location.

A single-pole off-and-on switch can be added in series with the collector of Q1 and the solid-state sounder to silence the sounder in case a bad bulb cannot be replaced immediately. (No need to be driven out of the car because of an annoying alarm sounder!)

The two burnout alarm circuits using a P.E. cell as the pick-up sensor function for almost any lamp but the car's headlights. There is no practical method in which to mount a P.E. cell to face and monitor the car's headlights. The problem is a mechanical one rather than an electronic circuit problem, so the solution must be an electronic one. The circuit in Fig. 2-5 will allow the monitoring of one or more lamps without the use of a P.E. cell.

The high-current lamp burn-out alarm circuit (Fig. 2-5) functions as a dc current sensing circuit. Transistor Q1, L1A, L1B, and associated component parts make up a Colpitts oscillator circuit. The frequency of oscillation is determined by the coils inductance and the value of capacitors C1 and C2.

With zero current flowing through L1B, the oscillator functions freely, producing a 5-volt peak-to-peak signal at the emitter of Q1. This ac signal is fed to a voltage doubler circuit (C4, D1, D2, and C5) that supplies a dc bias to the base of transistor Q2. A current-sensing control, R8, allows setting the circuit's sensitivity to detect currents from a low of 2 amps and up. As the current increases through L1B, the Q or efficiency factor of the oscillator's coil L1A decreases, causing a loss in the oscillator's output.

If the voltage at the base of Q2 drops below the preset emitter voltage, the LED and solid-state sounder remains inactive, but if a lamp burns out allowing the current flowing through the L1B winding to drop sufficiently or go to zero, the bias on Q2 rises, turning on the LED and sounder. You can set the sensitivity of the alarm circuit to detect the failure of a single lamp when two or more lamps are wired in parallel.

Hints on Building the High-Current Circuit

You can mount the majority of the circuit components on perf board with push pins, or make a p.c. board. Use whatever system or scheme suits you the best, because the circuit is not critical in its operation or parts layout.

The current sensing coil is wound on a 4-×-¼-inch ferrite rod. 75 turns of number 26 enamel-covered copper wire is wound at one end of the ferrite rod (Fig. 2-6) between two rubber grommets spaced 1¼ inches apart. Wind the coil in a solenoidal fashion and tape in place leaving about three inches of wire at each end of the coil to connect to the circuit.

Locate the current-carrying wire going to the lamp or lamps that are to be monitored and see if there is enough slack in the wire lead to wind a 4- to 8-turn coil on the end of the ferrite core. If you cannot wind L1B using the car's wiring, you can wind a coil using a length of #10 or #12 solid copper wire connected in series with the lamp current-carrying lead. Locate the alarm circuit as close to the lamp's wiring as possible, or if a remote location is required, be sure to use wiring that is heavy enough to carry the lamp's current.

The actual number of turns needed for L1B is determined by the current flowing in the lamp circuit. As the number of turns is increased on L1B the sensitivity of the circuit to lower current values is also increased.

If the lamp circuit wire is available, wind eight turns for L1B. The current-sensor circuit should then be usable for all applications. R8 gives a wide range of adjustment and allows for a wide variance in the number of turns used for the current-sensing winding, L1B.

Setting Up the Alarm Circuit

With the alarm circuit wired in place, turn on the lamp(s) that the circuit is monitoring, and set R8 to just where the LED and sounder cease to operate.

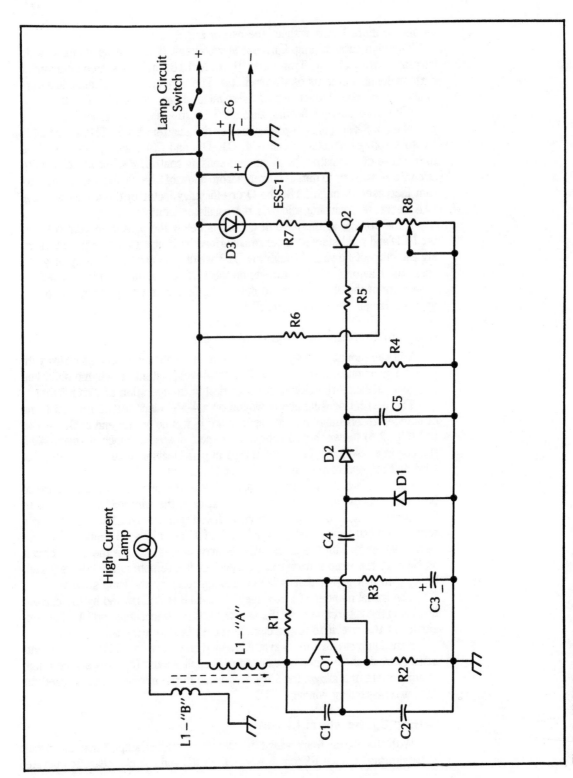

Fig. 2-5. (A) High current lamp burn-out alarm.

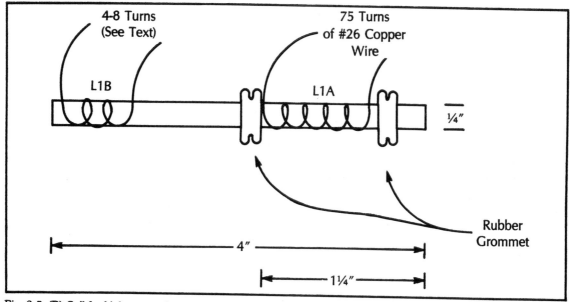

Fig. 2-5. (B) Coil for high current burn-out alarm circuit.

To test the alarm circuit, unplug any of the lamps in the circuit, and the alarm should sound off. If the lamp circuit contains only a single lamp, the adjustment of R8 can vary a great deal and still operate okay, but as the number of bulbs increases the adjustment becomes more critical.

Use the high-current lamp burn-out alarm circuit to keep an eye on the lamps that cannot readily have a P.E. cell mounted close to its light source. Use the P.E. cell sensor circuit for all other lamp-monitoring jobs.

Table 2-4. Parts List for Fig. 2-5.

D1, D2	1N914 silicon diodes
D3	LED
C1, C4	.12 mF 100Vdc mylar capacitor
C2, C5	.27 mF 100Vdc mylar capacitor
C3	6.8 mF 25Vdc electrolytic capacitor
C6	470 mF 25Vdc electrolytic capacitor
ESS-1	Solid-state sounder
L1-A	75 turns #26 enameled copper wire on ¼-inch ferrite rod
L1-B	4-8 turns on end of ferrite rod; use lamp wire if possible (see text)
R1	220 kΩ ¼ W resistor
R2, R3, R6, R7	1 kΩ ½ W resistor
R4	100 kΩ ¼ W resistor
R5	4.7 kΩ ¼ W resistor
R8	500 Ω pot
Q1, Q2	2N2222 npn general purpose silicon transistor
Misc.	4 x ¼-inch ferrite rod material, perf board, push-in pins, cabinet, wire, solder, etc.

MONITORING BRAKE FLUID

The car's brake system is as important, if not more so, as the battery used to start the car's engine. If the battery goes dead and the car cannot be started, usually another means of transportation can be arranged, but if an operating vehicle fails to stop on command, the results could prove harmful or even deadly to the car's driver and passengers.

Not all aspects of the brake system can be easily checked electronically, but the brake-fluid system can be monitored and a warning given if the fluid drops to a level that could cause a braking failure. The sensor circuit in Fig. 2-6 is probably the easiest method to use in keeping tabs on the brake-fluid level. All brands of brake fluid checked with an ohmmeter showed conductance and offered a resistance of about 20 kΩ between the meter's probe and the metal brake-fluid container. A circuit designed to monitor the resistance or conductance of the brake fluid is an inexpensive way to keep the fluid at a safe level. The circuit shown in Fig. 2-6 does so.

How the Circuit Operates

P1 and P2 are two small needle-like probes that extend through insulating seals into the brake fluid located in the dual master cylinder.

Resistors R1 and R2 supply bias for both transistors, Q1 and Q2. If the brake fluid is at a safe level, the majority of the bias current is passed on to ground through the fluid and neither of the transistors receives enough bias cur-

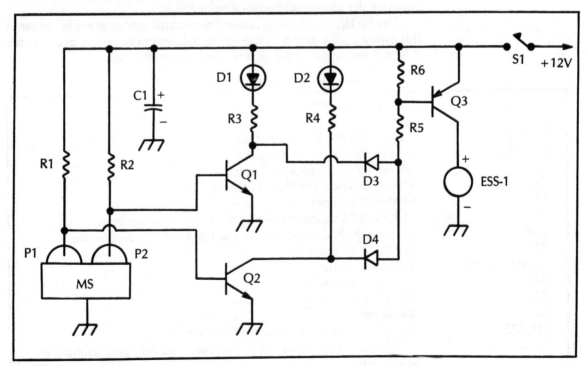

Fig. 2-6. Brake-fluid alarm circuit.

24

Table 2-5. Parts List for Fig. 2-6.

C1	100 mF 25V electrolytic capacitor
D1, D2	Red LED (any suitable color will do)
D3, D4	1N914 silicon signal diode
ESS-1	Solid-state electronic sounder
Q1, Q2	2N2222 npn general purpose transistor
Q3	2N3638 pnp general purpose transistor
R1, R2	(See text for determining values)
R3, R4	1 kΩ ½ W resistor
R5	4.7 kΩ ½ W resistor
R6	100 kΩ ½ W resistor
S1	SPST slide or toggle switch
Misc.	Two probes (see text), housing, perfboard, etc.

rent to turn on. All alarm indicators remain off. If either of the reservoir levels drops below the reach of the probes, the bias current will be passed into the base-emitter junction of Q1 or Q2, lighting one of the LEDs and supplying bias for Q3, thus sounding the solid-state alarm sounder. The resistor values for R1 and R2 might need to be a different, or a selected, value, because of the actual measured value of resistance between the probes and the brake fluid in the master cylinder.

Building and Using the Alarm Circuit

Again, for this circuit, you can use any construction scheme that works best for you and fits the application. The only semicritical area in building the project is selecting and mounting the two pick-up probes. If the master cylinder cover is made of plastic or some other nonconductive material, you can use good-quality sewing needles for the two probes. Clean the area on the cover where the needle probe passes through. Use a small amount of epoxy or silicone rubber and seal in place. The probe should be just deep enough to keep the alarm from sounding off when the fluid is sloshing around in the master cylinder while the car is in motion. After you find the best probe location, glue the probes in place with epoxy.

The two resistors, R1 and R2, need to be large enough to allow the voltage at the bases of Q1 and Q2 to be below the level needed to bias the transistors on. A good starting value for the resistors is 330 k. With the circuit in operation, measure the voltage at the base of each transistor. If the voltage is below .15 volts, the selected resistor value is okay, but if the voltage is higher, select a value that lowers the base voltage to .15 volts or below. When making this selection, be sure the two probes are in place, with at least ½ inch of brake fluid surrounding them.

After making the resistor selection, slowly remove the master cylinder cover and note when the alarm sounds off. If the alarm goes off too soon before the probes are halfway out of the brake fluid, the resistor values should be larger. You can determine the maximum resistance by using a resistance decade box. Connect the decade box as a substitute resistor for either R1 or R2.

Start with a value of 220 k, with the probe out of the fluid, and go up in value until the LED and sounder stop operation. This is one step above the maximum resistor that can be used for R1 and R2. Pick a resistor value for R1 and R2 that is between the upper and lower values determined by the voltage measurements and decade box.

Locate LEDs D1 and D2 for easy access so when the sounder goes off the LED indicates which vessel is low on fluid. If you don't want a visual indication, remove D1 and D2, and connect R3 and R4 directly to the plus bus line.

ANOTHER BRAKE-FLUID ALARM

The second brake-fluid level alarm circuit is a much simpler one than the first, and only works with the nonmetallic master cylinder fluid containers. Figure 2-7 shows the circuit of the second brake-fluid alarm. A floating boat rides the brake fluid with a small ferrite magnet attached. Position the magnet near a reed switch mounted on the container's lid. As long as the fluid level remains normal, the magnet keeps the reed switch activated and the LED (D1) remains lit, but the solid-state sounder stays off.

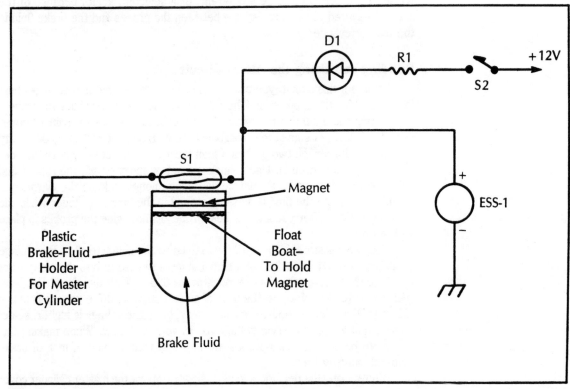

Fig. 2-7. Brake-fluid alarm circuit.

Table 2-6. Parts List for Fig. 2-7.

D1	Red LED (any suitable color will do)
ESS-1	Solid-state electronic sounder
R1	1 kΩ ½ W resistor
S1	Reed switch
S2	SPST toggle or slide switch
Misc.	Cabinet, wire, epoxy, etc.

If the fluid in the master cylinder drops to an unsafe level, the reed switch opens, allowing current to flow into the ESS-1 solid-state sounder and warning of a possible brake failure.

Building and Using the Simple Alarm Circuit

The electronic circuit is much too simple to offer any construction hints. Just be sure to do a neat wiring job. Almost any small plastic cabinet will do to house the few parts, and you can use a small terminal strip for all internal connections.

You can make the floating boat of any lightweight material (such as cork, wood, etc.) that floats and does not cause a reaction with the brake fluid. Make the floating boat by cutting a circle of material to just fit into the fluid's reservoir with a ¼-inch clearance around the edges. If the reservoir is not round, cut the boat to match the shape of the container. Mount a small, strong, lightweight ferrite magnet in the center and on top of the boat. Use either silicone rubber or epoxy, to permanently mount a small reed switch to the center and on top of the brake-fluid container.

No matter which brake fluid alarm circuit you build and use, a bit of patience mixed with the joy of experimenting goes a long way in making these a success and is an excellent approach to take when building any project.

TEMPERATURE SENSOR

Excess heat is a major enemy to a car's electrical and mechanical system, but if an overheating condition can be caught before it escalates to the danger point, the item in jeopardy can be saved, or at least removed before added damage occurs. The temperature sensor and alarm circuit in Fig. 2-8 can be used to monitor almost any component on an automobile.

The duo-diode temperature probe can be placed on any car part, and the circuit set to sound the alarm when the temperature rises too high. The circuit can also be used to keep tabs on the car's cooling system by locating the probe in the coolant liquid.

How the Temperature Alarm Works

A quarter of a LM324 quad op amp functions as high gain comparator circuit that compares the probe's voltage to a preset alarm limit voltage.

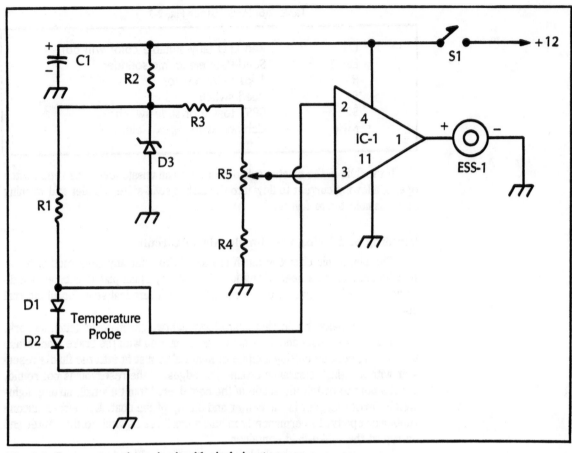

Fig. 2-8. Temperature alarm circuit with single input sensor.

Table 2-7. Parts List for Fig. 2-8.

C1	470 mF 25V electrolytic capacitor
D1, D2	1N914 silicon signal diode
D3	1N4734 5.6V zener diode
ESS-1	Electronic solid-state sounder
IC-1	LM324 quad op amp
R1	2.2 kΩ ¼ W 5 percent carbon film resistor
R2	1 kΩ ¼ W 5 percent carbon film resistor
R3	1.5 kΩ ¼ W 5 percent carbon film resistor
R4	270 Ω ¼ W 5 percent carbon film resistor
R5	500 Ω pot
S1	SPST toggle or slide switch
Misc.	Cabinet, perfboard, pins, wire, solder, etc.

Zener diode D3 offers a stable dc voltage for the input circuit of the comparator IC. R1 supplies a stable forward current to the two silicon diodes that make up the temperature probe. The voltage at the anode of D1 is about 1.2 volts at room temperature. As the temperature rises, the voltage across the diode pair goes down in a near linear fashion. The probe voltage is fed to one input of the comparator circuit, pin 2 of the IC-1. The reference voltage is taken from the wiper of pot R5 that presets the desired trigger temperature. As long as the voltage at the output of the probe remains higher than the reference voltage, the IC is turned off and no voltage appears at its output, pin 1. As soon as the probe's voltage drops below the reference voltage level, the IC switches on, supplying voltage to operate the solid-state sounder.

Building and Using the Temperature Alarm Circuit

The circuit is non-critical and it can be built to suit your own requirements. You can use perf board to mount the parts and a plastic cabinet to house it. The only component that requires some special attention is the temperature probe.

The two diodes making up the probe can be protected from environmental damage and insulated from any external voltage by covering them with a thin coat of epoxy. Try to keep the epoxy layer as thin as possible to allow the probe to respond quickly to any temperature change, as a too thick layer tends to isolate and delay the temperature change reaching the two diodes making the circuit respond sluggishly.

The alarm circuit can be preset to sound off at any temperature below 212° F by placing the probe into a container of water and heating it to the desired temperature, and setting R5 to activate the ESS-1 sounder. A standard glass thermometer can be used to monitor the water temperature while setting up the circuit's calibration.

The temperature probe can be mounted on most any car part that's susceptible to damage by overheating. A simple way to monitor the car's coolant system is to mount the probe directly on the radiator, but take care that neither of the probe's leads touches the metal and shorts out.

If more than one temperature-sensitive area needs monitoring the expanded circuitry in Fig. 2-9 allows up to four sensors to be used. The circuit is an expansion of the single input sensor circuit in Fig. 2-8, with a couple of circuit additions to make the expansion possible. Transistor Q1 is used as a buffer amplifier that helps isolate the loading effect of the four input circuits from affecting the reference voltage of the zener diode. Four LED diodes, D14-D17, visually indicate which of the sensors has been triggered when the solid-state sounder sounds off.

Use the same construction and calibration methods with the four-input circuit as with the single input sensor circuit, or choose whatever system or method works best for your application. With the number of diodes used in the expanded circuit, be sure to double-check each one to see that it is in the proper location and connected in the right polarity.

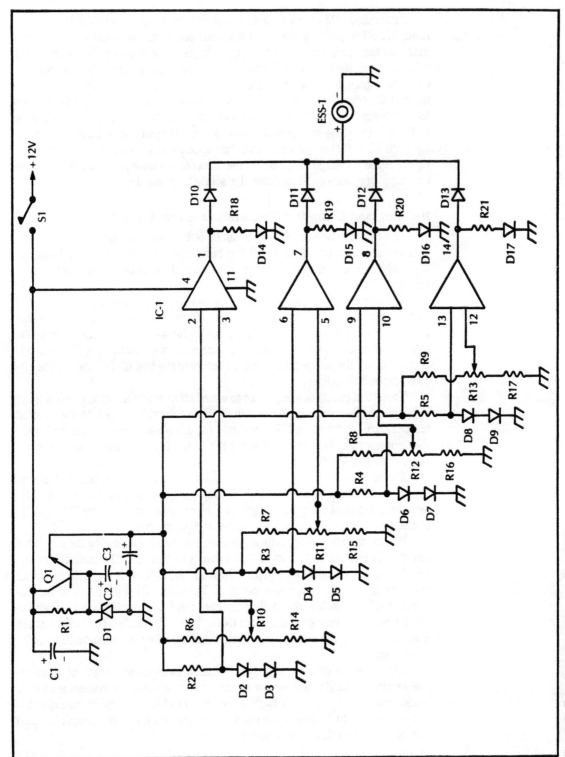

Fig. 2-9. Four input temperature alarm circuit.

Table 2-8. Parts List for Fig. 2-9.

C1	470 mF 25V electrolytic capacitor
C2, C3	100 mF 16V electrolytic capacitor
D1	1N4734 5.6V zener diode
D2-D13	1N914 silicon diode
D14-D17	LED
ESS-1	Electronic solid-state sounder
IC-1	LM324 Quad op amp
Q1	2N2222A npn general purpose transistor
R1, R18-R21	1 kΩ ½ W resistor
R6-R9	1.5 kΩ ½ W resistor
R10-R13	500 Ω pot
R14-R17	270 Ω ½ W resistor
S1	SPST toggle or slide switch
Misc.	Perfboard, pins, wire, etc.

A BACK-UP WARNING

If a vehicle is used in an area where children or pets are normally found, a back-up warning system is a must. The alarm should emit a warning signal that a kid who's engrossed in play cannot ignore, but don't rely on the back-up alarm alone—always look twice before moving a vehicle in any direction.

The circuit in Fig. 2-10 demands the attention of anyone in range of the dual bee-bop European-type warning sound each time the vehicle moves in a reverse direction. Two 555 timers are used in a dual oscillator circuit to produce the up-and-down frequency shift needed to obtain that unique warning sound. IC-1 and its associated components make up a very low frequency oscillator circuit that sets the up-down frequency shift rate, and operates at less than 1 Hz per second. IC-2 and its surrounding component parts produce an oscillator circuit that operates at several hundred cycles per second and drives the power transistor, Q1, which supplies the speaker with a high-level sound output. The slow up-down signal at pin 3 of IC-1 is fed into pin 5 of IC-2, giving it the dual tone output.

Building and Hooking Up the Back-Up Alarm

The circuit components can be mounted on a section of perf board, or a p.c. board can be made and used, but in either case, the layout isn't critical and any good construction scheme will do. The power transistor, Q1, requires a heat sink and a simple one can be made by taking a 3 × 3-inch piece of 1/16-inch aluminum material and mounting the transistor in the center. The heat sink and power transistor assembly must not be allowed to touch any other circuit parts, as a short here can blow the transistor or speaker, and maybe both.

The only time the circuit needs to be activated is when the vehicle's transmission is placed in the reverse position. A mechanically-activated switch is avail-

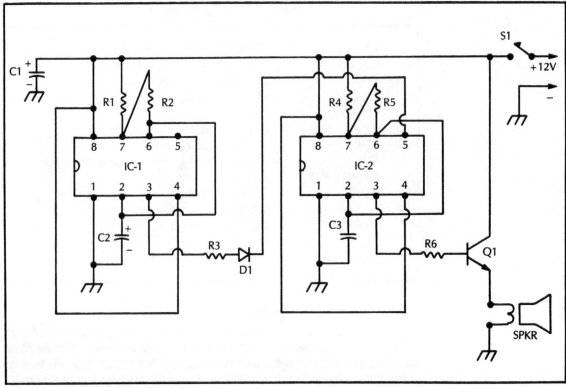

Fig. 2-10. Back-up alarm circuit.

able on some vehicles, but most often you will need to add a normally-open switch to the shifting mechanism. (Even when a switch is furnished, you may not be able to use it for the back-up alarm if it interferes with the original purpose of the switch, so it could be best to add your own switch.) Almost any type of switch can be used for S1 as long as its current rating is 3 amps or better. Don't try a magnetically operated dry reed switch for this application: the switch's contacts would be welded together the first time they made contact.

Experimenting with the Sounder Circuit

Now that you know about building, hooking up, and using the back-up alarm, it is time to offer a few simple circuit modifications that produce a number of different output sounds. If a slower up-down frequency shift rate is desired, increase the value of C2. Doubling the value of C2 cuts the frequency in half, and reducing the value of C2 by one-half doubles the up-down rate. To produce a higher output frequency, the value of C3 needs to be made smaller in value, and for a lower tone the value needs to be larger. Try going up or down in the value of C3 by 25 percent each time until the desired results are achieved.

The back-up alarm sounder circuit is not limited to this type of application, but can be used as is, or modified to operate as the main electronic sounder for a car or home burglar system.

Table 2-9. Parts List for Fig. 2-10.

C1	500 mF 25V electrolytic capacitor
C2	6.8 mF 25V electrolytic capacitor
C3	.12 mF 100V mylar capacitor
D1	1N914 silicon diode
IC-1, IC-2	555 timer IC
Q1	2N3055 npn power transistor
R1, R2	100 kΩ ¼ W resistor
R3, R6	1 kΩ ¼ W resistor
R4	2.2 kΩ ¼ W resistor
R5	10 kΩ ¼ W resistor
SPKR	16 Ω outdoor metal horn-type speaker
S1	Switch activated by the reverse shift mechanism
Misc.	Perfboard, pins, wire, cabinet, etc.

A BURGLAR ALARM FOR THE CAR

Burglar alarm systems designed especially for automobile use are similar to home alarm systems, with only one small difference: in the vehicle, the power source is limited to the 12-volt battery. For most mobile alarm applications this limitation does not cause any difficulties in the design or working quality of the system. The more unorthodox or non-standard an alarm system becomes in its design, the more difficult it is for an experienced burglar to circumvent the system and make a safe entry. In many instances, a good home-designed alarm system causes even the expert burglar to pass and go on to easier pickings. Don't think building your own is going to give you a second rate system, because a good home-built system can be the most difficult to penetrate, as there is no history or factory information on your one-of-a-kind alarm system.

A simple but versatile car burglar alarm system is shown in Fig. 2-11. Any number of input sensors can be used with the circuit by adding extra input diodes (D1-D3), and the output SCR switching circuit can be used to activate any alarm sounder units including the circuit in Fig. 2-10.

The success of this alarm circuit is in the operation and location of the function control switch, S1, and the selection of input sensors. The location of S1 should be well hidden from anyone not authorized to use the vehicle and the trip switches that activate the alarm's input should be selected to best fit the type of automobile the system is used in.

To better understand the alarm's functions and limitations, a complete explanation of the alarm's circuit operation is in order. With S1 in the on position, and with no plus voltage present at any of the inputs (A, B, and C), the circuit is on and at rest. The voltage at the base of Q1, the emitter of Q2, and the gate of Q3 is zero. All three of the semiconductors are in the non-conduction state.

If one or more of the input sensors (A, B, or C) is activated with 12 volts, the electrolytic capacitor C1 begins to charge through the timing resistor, R1.

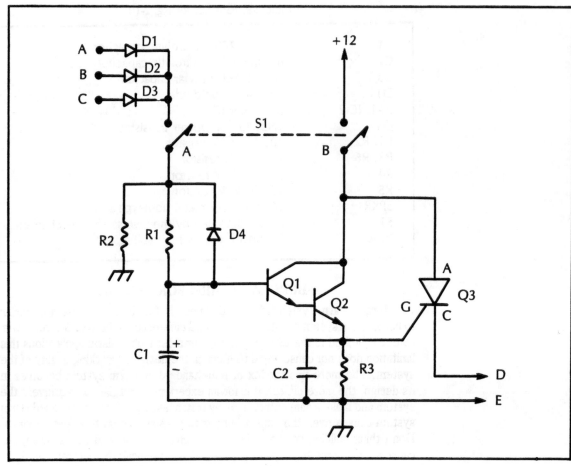

Fig. 2-11. Car burglar alarm circuit.

Table 2-10. Parts List for Fig. 2-11.

D1-D4	1N4002 silicon diode
C1	47 mF 16V electrolytic capacitor
C2	.1 mF 100V ceramic disc capacitor
Q1, Q2	2N2222 npn general purpose transistor
Q3	C103 SCR or any similar general purpose SCR
R1	680 kΩ to 1.5 M Ω ¼ W resistor (see text)
R2	27 kΩ ¼ W resistor
R3	2.2 kΩ ¼ W resistor
S1	DPST toggle or slide switch
Misc.	Terminal strip, cabinet, wire, etc.

Transistors Q1 and Q2 are connected in a Darlington high input impedance circuit that isolates the timing capacitor C1 from any loading.

As long as the input voltage is present, the voltage level continues to increase across C1 and at the emitter of Q2. When the voltage reaches approximately .75 volts at the gate of Q3, it turns on and activates the alarm sounder connected to the D and E terminals. A 680 k timing resistor (R1) delays the alarm by about 10 seconds and a 1 Meg resistor offers about a 15 second delay. Keep R1's value below 10 MΩ which, if the leakage quality of C1 is good, offers a delay of about three minutes. Another way to change the delay time is to change the value of C1. Doubling C1 to 100 μF just about doubles the delay, and cutting the value of C1 in half reduces the delay by about the same amount. By experimenting with these two component values, almost any desired time delay can be obtained.

A number of the car's electrically operated items can be used as sensor switches for the alarm's input circuits (A, B, and C). The only requirement is that when a switch is activated that turns on some electrical item, a plus 12 volts must be switched on to that electrical device. The alarm's inputs (A, B, and C) can be tied on to any of these electrically-switched items.

The car's dome light, trunk light, headlights, ignition coil, radio, or tape player can be used for sensor switches as long as each meets the requirements of a switched plus 12 volts.

If a sensor is activated for only a short period of time, the function of D4 and R2 is to quickly discharge C1 when the 12 volts is removed from the sensor's input. This circuit keeps the actual set time delay more accurate in its complete timing cycle, but both parts can be removed if the circuit feature is not needed. The SCR can be as small as the 1 amp device shown or any higher-current SCR can be used to meet whatever current requirement is needed for the alarm sounder.

In some cars, several electrical items use a mechanical switch to apply power to the negative side of the battery and not to the positive as is required by the alarm circuit. The circuit shown in Fig. 2-12 is a voltage inverter circuit that allows any negatively switched circuit to be used with the alarm unit.

The circuit shows the inverter circuit connected to a car's door switch that uses a grounding switch to light the lamp. With the car door closed, the lamp is off and the transistor Q1, in Fig. 2-12, is biased on through R1 and the dome lamp. The voltage at the collector is near zero and no signal is sent to the alarm's input. When the door opens, the lamp lights and the bias is removed from the base of Q1 letting its collector voltage rise to nearly 12 volts. This plus voltage is fed through one of the input diodes, D1-D3, starting the timing capacitor charging. Any number of inverters can be built as is needed for the installation; just repeat the circuit as shown in Fig. 2-12.

Building and Using the Alarm Circuit

Here again is a noncritical circuit. Any good construction scheme will do. The main trick is to locate S1 in an accessible but not obvious place where it can be turned off and on as you enter and leave the car. Select from one to

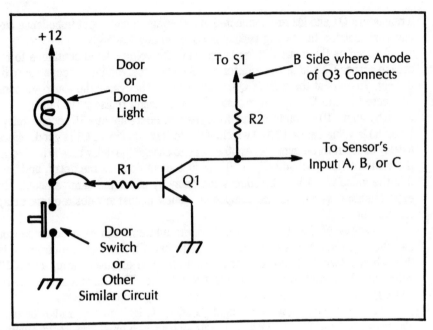

Fig. 2-12. Voltage inverter circuit.

Table 2-11. Parts List for Fig. 2-12.

Q1	2N2222A npn general purpose transistor
R1	10 kΩ ¼ W resistor
R2	4.7 kΩ ¼ W resistor
Misc.	Perfboard, pins, wire, etc.

three electrical circuits to connect the sensors to, and use the inverter circuit if needed.

Knowing when to turn S1 on and off is easy if you just remember that as long as you are in the car keep S1 turned off, and when you are out of the car have S1 turned on, arming the alarm circuit. If the timing circuit is selected for a 15 second delay then you have that amount of time to turn off the alarm after you have entered the car, and the same amount of time to exit the car after the alarm has been armed.

Chapter 3
Intrusion
Alarm Sensors

No matter how simple or complicated an alarm system might be, the single most important component is the input sensors that tell the master control unit when a break in security has occurred and the location of the breach. The most common and frequently used input sensor is the simple mechanically activated switch that is either normally open or normally closed in the set condition.

At least two normally closed circuit alarm sensors are one-time-use-only sensors, destroyed when they are set off.

The first, and most simple, of the two sensors is the break wire sensor that is no more than a thin piece of wire stretched across an area where an intruder would break it upon making an entry. The second sensor is one we've all seen at one time or another on store windows and doors; strips of lead tape attached to each of the glass panes.

The break wire sensor can be upgraded one step to remove it from the one-time-use-only system to a reusable sensor. That is accomplished by using a larger wire with a small plug attached to both ends. Each end of the trip wires plugs into a quick-disconnect type jack so when the wire is disturbed it pulls free from one or both of the jacks setting off the alarm.

A BASIC ALARM CIRCUIT

The circuit shown in Fig. 3-1 is about as elementary as a burglar alarm system can be while still offering a fair degree of security. In fact this very alarm circuit has been a mainstay in burglar alarms used for over a half century. Many are still in use protecting homes, stores, factories, and other types of valuables.

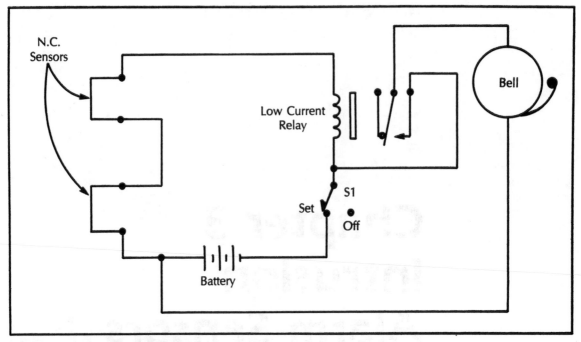

Fig. 3-1. Elementary burglar alarm system.

The alarm's circuit operates in the following manner: with all of the N.C. (normally closed) sensors intact and the function switch S1 placed in the set position, the relay is energized, as shown in Fig. 3-1, and the relay's contacts connecting the bell to the battery are held open. Let any one of the sensors break, or cut the wire leading to the sensor circuit, and the relay falls out, setting the bell to ringing.

The circuit shown in Fig. 3-2 is designed to use N.O. (normally open) sensors to activate the alarm. This alarm circuit has a very definite advantage over the previous circuit, because the battery is not used until the alarm has been set off. The circuit is not the ultimate in burglar alarm circuits. There is a small fault in its design: if the N.O. sensor switch is only momentarily closed and reopened, the alarm bell only rings for that short period. Not too good if the burglar is fast!

To overcome the circuits limitation, a simple modification is made as shown in Fig. 3-3. This turns the relay into a self-latching type that keeps the bell ringing until the battery goes dead or the circuit is turned off.

A closer look at the circuit reveals that the zero standby current remains a feature, and the second set of relay contacts keeps the relay energized even if the N.O. sensor switch is momentarily closed and then re-opened. So why isn't this the ideal alarm circuit to use? If either or both of the sensor leads are cut the alarm is completely disabled and does not make a sound. Standing alone, the N.O. type sensor is the least secure type to use, but when the N.O. and the N.C. sensors are mixed in a single alarm system a very secure burglar determent system can emerge.

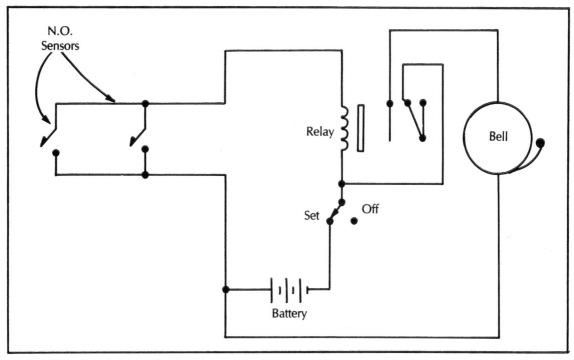

Fig. 3-2. Simple N.O. sensor burglar alarm system.

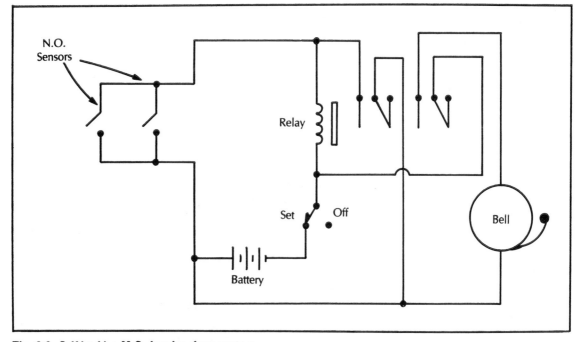

Fig. 3-3. Self-latching N.O. burglar alarm system.

The alarm sensor circuits offered in this chapter are designed to work with any alarm control system that you may now be using, or intending to use, that operates with either or both N.O. and N.C. input sensor circuits. This includes 99% of all burglar alarm systems in use today.

A SINGLE CELL LIGHT PICK-UP CIRCUIT

If you could afford to hire a full-time security guard to keep a careful eye on your home or business for 24 hours a day there is still no absolute guarantee that the location would be 100 percent secure. Why? Because humans are not infallible.

A person placed on a long term sentry duty is exposed to boredom, mental fatigue, and other human frailties that tend to reduce their overall job efficiency. An equivalent electronic sentry is tireless in its duty and does not fail from any human weakness. The human sentry uses his eyesight and hearing to pick up on any unusual movement or sound and his brain in determining what action to take.

In a simple, but effective, burglar alarm system, photo cells are substituted in place of the human eyes. Photo cells never blink or sleep or look in the wrong direction, but remain on duty for 24 hours a day every day of the week. The positive feature of the simple photo cell is not without its limits when compared to the complex human eye, but if used properly does an excellent job in keeping an area under a visual and constant surveillance. Our first electronic sensor uses a photo cell to detect motion within a controlled area.

A single cell light pick-up circuit, shown in Fig. 3-4, is designed to watch an area between the photo cell and a light source.

The light pick-up sensor is made up of a P.E. cell mounted in one end of a length of opaque tubing with the front of the cell facing out through the full length of the enclosure. The light sensor is aimed toward a fixed light source that is located several feet away. With a clear area between the light and sensor, the sensitivity is adjusted with R9 to produce a plus 12-volts output at terminal A, and a closed circuit output at terminal B and C.

As long as the light path between the source and the sensor is unbroken, the two output circuit conditions remain unchanged. If any object comes between the two that casts a shadow on the P.E. cell, the outputs go into circuit reverse sending out an alarm signal. A large area can be protected with only a single light source located in the middle of the area covered, and a number of light sensors circuits aimed toward the source from different locations around the area. Actually, any existing light source can be used as long as a direct path exists between the source, the protected area, and the pick-up sensor.

Here's How the Circuit Operates

Transistors Q1 and Q2 are connected in a high input impedance Darlington amplifier circuit that offers almost no loading to the photo cell's output. The voltage at the base of Q1 is set to slightly less than half of the supply voltage with the circuit's sensitivity control pot R9. A fixed reference voltage is set

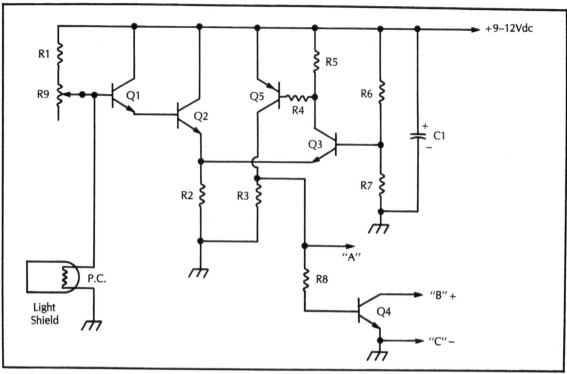

Fig. 3-4. Single-cell light pick-up alarm sensor.

at the base of Q3 with resistors R6 and R7 to produce a voltage of about 60 percent of the supply voltage. With the sensor aimed at a light source, the voltage at the base of Q1 is lower than the reference voltage set by the voltage divider, R6 and R7. This places the emitter voltage of Q3 below its base voltage, turning it and Q5 on and supplying a positive output at terminal A, and

Table 3-1. Parts List for Fig. 3-4.

C1	100 mF 25V
P.C.	Resistive type photocell Clairex #CL603A or similar
Q1-Q4	2N2222 npn silicon general purpose transistor
Q5	2N3638 pnp silicon general purpose transistor
R1	100 kΩ ¼ W resistor
R2, R4	4.7 kΩ ¼ W resistor
R3	3.3 kΩ ¼ W resistor
R5, R8	10 kΩ ¼ W resistor
R6	1.5 kΩ ¼ W resistor
R7	2.2 kΩ ¼ W resistor
R9	1 MΩ pot
Misc.	⅜-inch metal tubing, perfboard, pins, cabinet, light source, wire, etc.

a normal closed circuit output at B and C. Interrupt the light reaching the P.E. cell, and the voltage at the base of Q1 and emitter of Q2 rises to near the source voltage. This places the voltage at the emitter of Q3 higher than its base voltage, turning it and Q5 off and setting the output in an alarm condition.

Building the Light Sensor

The electronic component parts can all be mounted on a piece of perf board and housed in a metal or plastic cabinet. No special wiring scheme is required, but a neat job always pays off.

Figure 3-5 is a drawing showing how the light pick-up sensor is constructed. The length of the tubing helps to determine how far away the sensor can be located from the light source. As the length of the tubing is increased, the field of view through the tube is reduced, allowing the distance from the light source to be greater. The reduced field of view helps to keep out any other light coming from the area around the main light source. If too much secondary light hits the photo cell, it could reduce the circuit's sensitivity to medium or small objects.

For most sensor applications the tubing length can be 6 inches for short to medium distances, and 12 inches for long range use; here a little experimenting with different lengths of tubing will pay off in selecting the best length for a specific job. The sensitivity of the P.E. cell and the intensity of the main light source used are also factors that determine the maximum usable range of the sensor.

After a tubing length is selected, slide the photo cell in one end of the tubing as in Fig. 3-5. Insulate both of the cell's wire leads with spaghetti tubing, and seal the cell in place with opaque epoxy.

If the photo cell is too loose in the tubing, shim the cell with electrical tape or paper to guarantee that the cell is in proper alignment with the front of the tube. If the cell looks off to one side in a long length of tubing, the sensitivity is reduced, so a little extra care here helps make a better sensor.

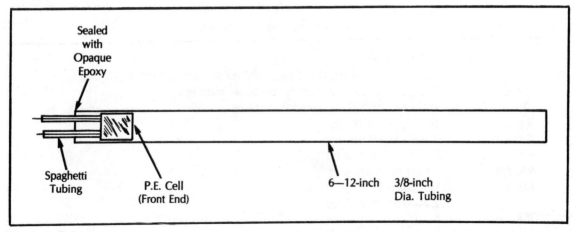

Fig. 3-5. Pick-up sensor.

Setting Up the Light Sensor for Use with an Alarm System

With a 9-volt power source connected to the sensor circuit and the tubing aimed toward a light source, connect a dc voltmeter to the emitter of Q1 and adjust R9 for a voltage of 3 to 4 volts. Connect the meter to terminal A and the voltage should be near 9 volts. Leave the meter connected here for the following sensitivity tests. While watching the voltmeter, walk between the light source and the sensor. The voltage should drop to near zero and remain there until the light once again hits the sensor.

Experiment with: a different light source; location of the sensor; the setting of sensitivity control R9; and the size of objects used to break the light beam. This will show you how to obtain the best results with the circuit.

The multi-input light sensor circuit in Fig. 3-6 allows up to five separate input sensors to be used at the same time and gives out a single N.C. circuit condition for all input circuits. The sensor's output circuit is the same as the one used in the circuit in Fig. 3-4, and can be used in the same way. Each of the input P.E. cells is housed in a section of opaque tubing as shown in Fig. 3-5, and can be as long as needed for each application.

Placing the five-input sensor circuit in service is similar to setting up the single input circuit with the following few variations in the set-up procedure. With power applied to the circuit: open switches S2 through S5, leaving only S1 in the on position (see schematic diagram in Fig. 3-6). Connect the positive lead of a high impedance dc voltmeter to the input of inverter 1, pin 3, and the negative lead to the circuit common or negative supply. Set the meter's range switch to the lowest setting that will read the circuits power source voltage without overloading the meter. Set R1 so the maximum resistance is in the circuit, and aim PC-1 toward a light source.

Move the P.E. cell assembly around while aiming it toward the light source, until the maximum voltage reading is obtained at pin 3 of the IC. This position indicates that the pick-up is aimed at the brightest part of the light source and is in the best possible location. The voltmeter should read at least ⅔ of the supply voltage at the input of the sensor for proper circuit operation. If a strong light source produces a voltage near to that of the power source, then adjust R1 for a voltage reading of ⅔ or ¾ of the supply voltage at the input of the inverter. By setting the input voltage to this level, the circuit will respond to a smaller object passing between the sensor and light source. The voltage at the output of inverter 6, at pin 15, should measure just slightly less than the circuit's supply voltage. The positive bias voltage at the output of inverter 6 keeps Q1 turned on, supplying a normally closed circuit at its output terminals B and C.

Follow the same set-up scheme for the remaining sensor inputs. Remember to turn on each of the inverter output switches as you go.

If the P.E. cell sensors are to be located at a distance of five feet or more from the main circuit, they should be connected with a length of two-wire shielded mike cable to reduce any possibility of the high input impedance circuit picking up noise or rf interference.

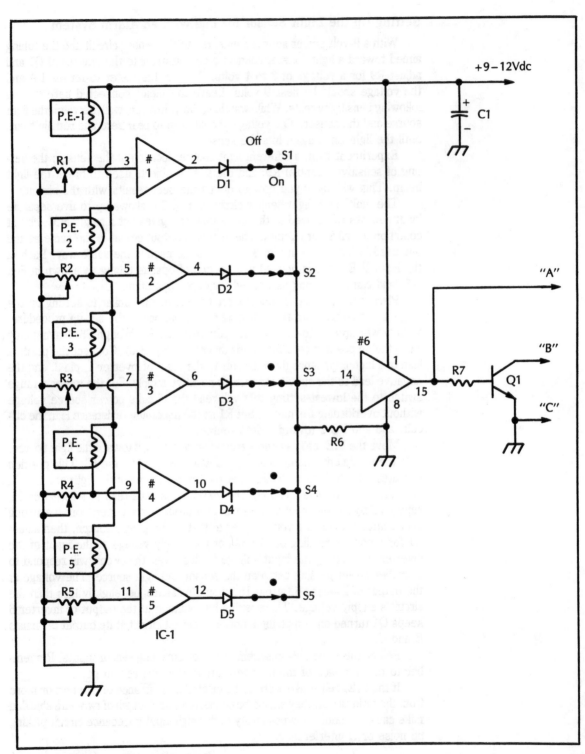

Fig. 3-6. Multi-input light sensor circuit.

Table 3-2. Parts List for Fig. 3-6.

C1	100 mF 25V electrolytic capacitor
PC1-5	Resistive type photocell, Clairex #C1603A or similar
Q1	2N2222 npn general purpose transistor
IC-1	4049 inverting hex buffer, digital (CMOS) IC
D1-5	1N914 silicon signal diode
R1-5	1 MΩ pot
R6	10 kΩ ¼ W resistor
R7	2.2 kΩ ¼ W resistor
S1-S5	STSP toggle or slide switch
Misc.	Perfboard, ⅜-inch tubing, wire, solder, etc.

If it is not possible to use the solid-state N.C. switching circuit, a small sensitive relay can be connected between terminal B and the positive power source. Also by using a DPDT relay, a choice of either N.O. or N.C. contacts can be used to work with any alarm system you might have available. The output at terminal A supplies a plus voltage near that of the power source used and can be used to drive up to 20 mA into an external circuit.

The construction of the multi-input sensor circuit can follow the same basic scheme as the single sensor circuit. No matter what construction method you choose to use, be extremely careful in handling the CMOS IC, as it can be damaged by static electricity. If the circuit is built on perf board with push-in pins, a 16-pin IC socket should be used for the CMOS IC.

CHOPPING THE LIGHT SOURCE

In a number of alarm installations, a special pick-up sensor is needed that can be used in total darkness or bright sunlight, without its presence being obvious to an intruder. Several different schemes come to mind that could fulfill this requirement, and the first one uses an invisible light source and detector to outsmart the intruder. You can add a new dimension to your alarm system by shooting an invisible light beam down a hall or across a room where an intruder would be likely to pass, breaking the light connection and setting of the alarm.

The simplest method (from the standpoint of the circuit design) in any light-operated alarm sensor is a continuous beam of light that keeps the detector activated and ready for a break in the light source. Ninety-nine percent of the time this method works very well, and can be designed and installed where it is nearly impossible for an intruder to breach. But in the remaining 1 percent, where an expert burglar might know either from experience or inside information that he could aim a portable light source at the detector and bypass the sensor, a slightly more sophisticated sensor system is needed.

One method that can be used to increase the security of the sensor is to chop the light beam up into small parts at a low frequency rate and tune the detector circuit to see only this ac light source.

The circuit in Fig. 3-7 performs half of the job by chopping up an invisible infrared light source at a rate of 1500 Hz, and aiming it toward the infrared detector, Q1, in the receiver circuit shown in Fig. 3-8.

Here's How the Circuit Operates

Starting out with the light transmitter circuit shown in Fig. 3-7, a 555 IC is connected in a free-running oscillator circuit operating at a frequency of 1500 Hz with the output switching the infrared diode off and on at that rate. The frequency of the 555 oscillator is determined by the components values of R2, R3,

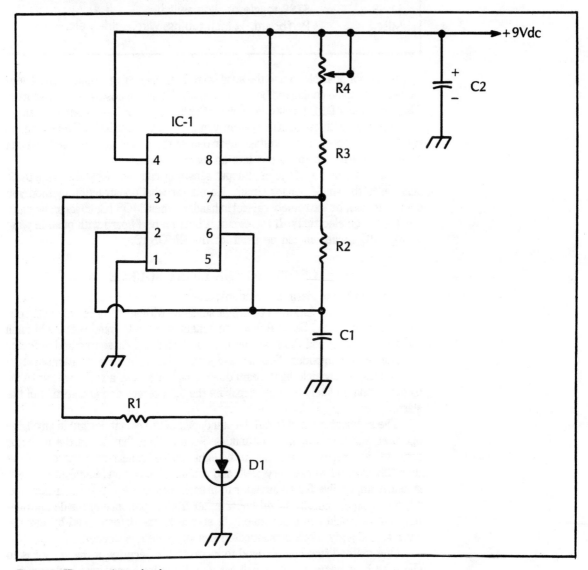

Fig. 3-7. IR transmitter circuit.

Table 3-3. Parts List for Fig. 3-7.

C1	.068 mF 100V mylar capacitor
C2	100 mF 16V electrolytic capacitor
D1	High output infrared LED #XC-880-A (Radio Shack #276-143)
IC-1	555 timer IC
R1	270 Ω ½ W resistor
R2	4.7 kΩ ¼ W resistor
R3	2.2 kΩ ¼ W resistor
R4	10 kΩ pot
Misc.	Perfboard, cabinet, wire, 9V power source, etc.

R4, and C1, with the pot R4 setting the exact frequency to match the tuning of the receiver. The exact operating frequency of the sensor isn't too important, as long as both the receiver and transmitter are operating on the same frequency. The resistor value of R1 determines the drive current for the diode and if the resistance value is changed to increase the light output, be sure to check the maximum allowable power dissipation for the infrared diode that you use in the circuit.

The receiver circuit shown in Fig. 3-8 is somewhat more involved than the simple transmitter circuit, but as receivers go, it too would be classified as a simple circuit.

An infrared light detector transistor Q1 intercepts the pulsed transmitter signal and feeds the small ac signal to the input of a twin-T filter circuit. A close look at the filter circuit shows that its input and output are connected between the base and collector of Q2. This is a notch twin-T filter which offers the maximum attenuation between its input and output at its tuned frequency. At the notch frequency the twin-T impedance is at its maximum, which offers the very minimum of negative feedback for the transistor amplifier stage (Q2). The gain of the amplifier is at its maximum under these circuit conditions, giving the greatest output at the center or notch frequency. In brief, this is how the active filter circuit operates to see and amplify only the 1500 Hz chopped light signal coming from the sensor's transmitter. Transistor Q3 adds voltage gain to the 1500 Hz signal to raise it to a level great enough to be rectified by the voltage doubler, D1 and D2, and to bias Q4 on giving a closed circuit output at terminals B and C.

Building the Light Chopper Sensor

Any good construction scheme will do fine, but each of the infrared diodes should be mounted where ambient light doesn't hit the light-sensitive and emitting surface of the device.

This can best be accomplished by locating the IR devices inside the opaque cabinet that houses each of the circuit components, as shown in Fig. 3-9. Drill or punch a ⅜-inch hole in one end of the cabinet and mount the IR device about

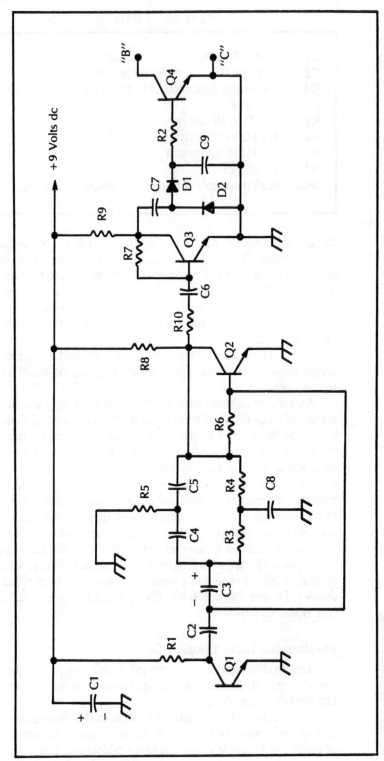

Fig. 3-8. IR receiver circuit.

Table 3-4. Parts List for Fig. 3-8.

C1	100 mF 16V electrolytic capacitor
C2, C9	.22 mF 100V mylar capacitor
C3	4.7 mF 25V electrolytic capacitor
C4, C5, C6, C7	.1 mF 100V mylar capacitor
C8	.2 mF 100V mylar capacitor
D1, D2	1N914 silicon diode
R1, R2	10 kΩ ¼ W resistor
R3, R4	1 kΩ ¼ W resistor
R5	470 Ω ¼ W resistor
R6, R7	220 kΩ ¼ W resistor
R8, R9	2.2 kΩ ¼ W resistor
R10	270 Ω ¼ W resistor
Q1	Infrared photodetector (Radio Shack #276-142) or similar
Q2, Q3, Q4	2N2222 npn silicon general purpose transistor
Misc.	Perfboard, pins, cabinet, 9V power source, etc.

1 inch inside and in line with the hole. This method of mounting the IR transistor Q1 in the receiver's cabinet, helps keep the ambient light from swamping out the transmitter's signal. Mounting the transmitter diode in the same way adds to the system's directivity and security.

Using the IR Chopper Sensor

Choose a location a burglar would be likely to cross or pass through and locate the IR chopper sensor transmitter and receiver to cover that area. The chopper circuit works best if the two units are not separated by more than 15 feet, and some experimenting is necessary to find the best location for each of the sensors.

After the desired location for the transmitter and receiver is found, connect power to each circuit and connect a dc voltmeter to the receiver's circuit. The positive meter lead to the cathode of D1 and the negative lead goes to battery negative or circuit common.

While watching the meter, adjust R4 on the transmitter's circuit for a maximum meter reading, which should be at least 1½ volts and can be as much as 5 volts depending on the distance between the two units.

Terminals B and C of the receiver can be used with an alarm system that is set up to accept a solid-state N.C. sensor input circuit, or a low-cost dc relay can be connected between terminal B and the plus supply to operate with almost any alarm system in existence.

No matter which light sensor circuit you choose to use in your alarm system, the importance of a properly placed sensor cannot be overemphasized. A poorly placed sensor cannot only miss an intruder, but can cause a number of false alarms. There's no other single response which reduces the overall ef-

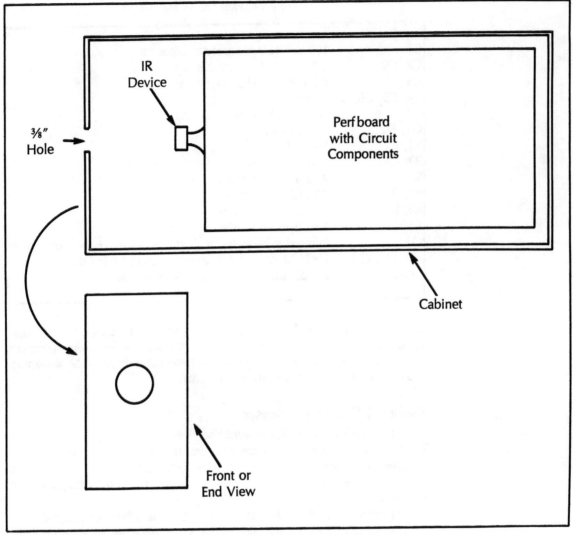

Fig. 3-9. IR mounted in cabinet.

fectiveness of any alarm system as a false alarm—call wolf too many times falsely, and no one pays attention to the alarm.

PROXIMITY ALARM SENSORS

I can't think of a more interesting alarm project to build, experiment with, and to place in service than the almost magical proximity alarm sensor. Here's a type of sensor that operates without any visible energy source using neither light or sound to protect a valuable object or cover a given area. The majority of the proximity sensor circuits operate by supplying an rf energy source to a metal pick-up plate and metering the energy losses as objects come near to or make contact with the sensor plate.

The block diagram in Fig. 3-10 illustrates a proximity alarm that operates as follows. A low rf frequency oscillator circuit is loose coupled to a L/C tank circuit, with a metal sensor pick-up plate attached at the junction of two small value coupling capacitors that complete the oscillator's feedback circuit. When an object moves close to the pick-up plate, a small amount of the rf energy is bypassed to ground lowering the oscillators output at the detector. If the object is large in size or very close to the pick-up plate, the oscillator ceases operation and remains inactive until the object is removed from the pick-up area.

Another proximity alarm sensor is shown in Fig. 3-11. This uses an R/C oscillator circuit in place of the L/C tuned circuit. The operation of the R/C proximity circuit is the same as that of the L/C circuit with the advantage of using a lower cost tuned circuit. An L/C and R/C proximity sensor project follows with a complete explanation of the circuit operation of each.

A third and most unusual proximity alarm sensor is shown in the block diagram of Fig. 3-12. A low power rf oscillator feeds an antenna transmitting plate that is placed facing a similar plate hooked to a receiver and level detector circuit. When an object passes between the two plates, a portion of the rf signal is directed to the object and ground, reducing the output at the receiver and detector circuit. The above three are just a few of the many types of proximity sensors that can be built and used with a successful burglar alarm system.

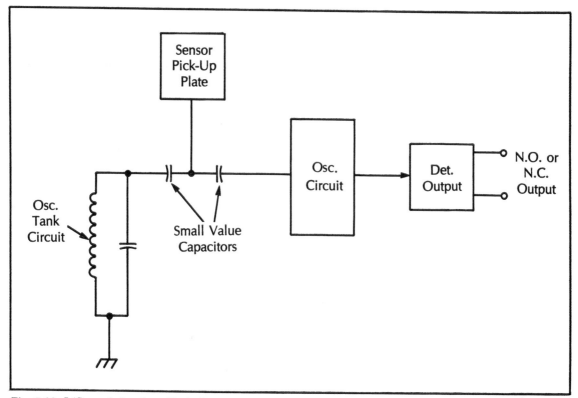

Fig. 3-10. L/C proximity alarm block diagram.

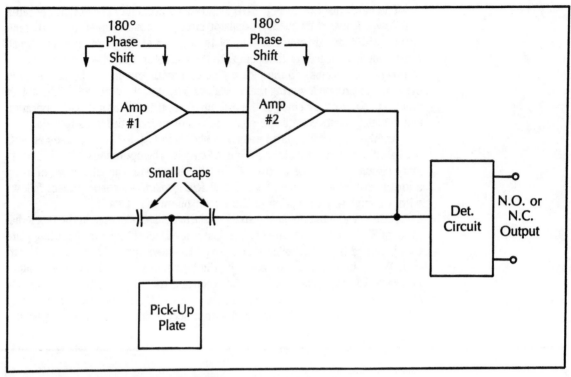

Fig. 3-11. R/C proximity alarm block diagram.

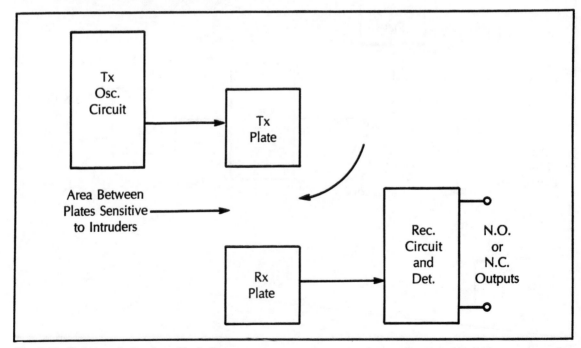

Fig. 3-12. Tx/rx proximity alarm block diagram.

Building a Sensitive L/C Proximity Sensor

An excellent proximity sensor circuit is shown in Fig. 3-13 that's designed to protect metal objects, such as file cabinets, safes, equipment, and other similar items.

Two high-gain NPN transistors, Q1 and Q2, are connected in a high input impedance Darlington amplifier circuit that's connected to a L/C tuned circuit. The combination functions as a low frequency Colpitts oscillator operating at approximately 30 kHz. The oscillator's output is coupled from the top of R4 to a voltage doubler circuit that supplies a foreword bias to transistor Q3 producing a N.C. output for the sensor.

The amount of feedback for the oscillator circuit is controlled by the setting of R4, and is used as a pick-up sensitivity control. The circuit's wide sensitivity adjustment range allows the sensor to protect an object as small as a rare coin or a safe the size of an elephant with a single setting of R4. Any regulated power source that can provide a stable output of 9-15 Vdc at a maximum of 20 mA will do just fine as power for the circuit.

Putting the Parts Together

A metal or plastic cabinet will do for housing the circuit and the components can be mounted on perf board with push-in pins, but any good construction scheme is fine. The L of the L/C tuned circuit is made by winding 75 turns of #22 copper wire on a ⅜ × 3 inch piece of ferrite rod, as shown in Fig. 3-14.

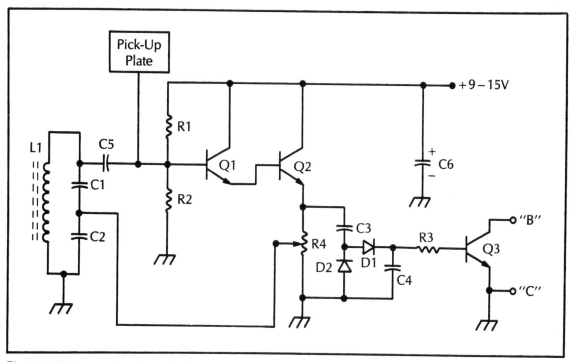

Fig. 3-13. L/C proximity alarm circuit.

Table 3-5. Parts List for Fig. 3-13.

C1, C2, C3	.068 mF 100V mylar capacitor
C4	.27 mF 100V mylar capacitor
C5	300 pF 100V mica capacitor
C6	100 mF 25V electrolytic capacitor
D1, D2	1N914 silicon diode
Q1, Q2	2N5089 npn high gain silicon transistor
Q3	2N3903 npn general purpose silicon transistor
L1	75 turns of number 22 copper wire on a ferrite rod ⅜-inches × 3 inches (see text for details)
R1, R2	1 MΩ ¼ W resistor
R3	4.7 kΩ ¼ W resistor
R4	1 kΩ linear pot
PICK-UP PLATE	See text
Misc.	Perf board, metal or plastic cabinet, 9-15 V power source, material for different pick-up plates, etc.

Wind the coil in layers in a solenoidal manner covering a 1-inch area in the center of the ferrite rod material. After 75 turns has been wound on the core, leave at least three inches of wire at each end of the winding for hook-up purposes and cover the winding with electrical tape. If a metal cabinet is used to house the circuitry, don't allow the ferrite coil to be any closer than 1 inch from any part of the metal cabinet. If the coil is located too close to the metal, the coil's Q will be lowered and could cause the circuit to malfunction.

Checking Out the Sensor

If an earth ground is handy, connect the sensor circuit's negative or common ground to earth ground. Connect the sensor's pick-up input to the object

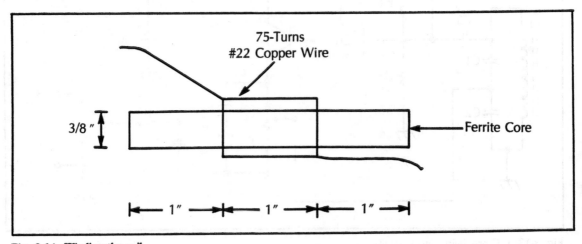

Fig. 3-14. Winding the coil.

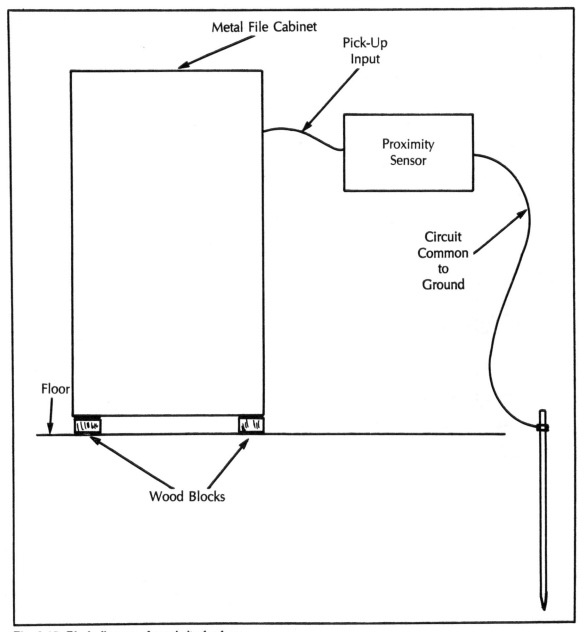

Fig. 3-15. Block diagram of proximity hook-up.

to be protected (see Fig. 3-15). The protected object must be isolated from the ground or floor with hard wood blocks or other excellent insulating material. The item should be elevated at least one inch above a concrete floor for the circuit to operate properly. Some experimenting here will help to find the best overall arrangement of the protected object and the circuit's sensitivity adjustment.

Connect the positive lead of a dc voltmeter to the circuit positive (top of C6) and the negative lead to terminal B (collector of Q3) and set the meter's range to read the voltage supplying the circuit.

While keeping a good distance from the protected object slowly turn R4 to the point where the meter goes from zero to a voltage reading near that of the supply. This is the adjustment area of R4 that will be used for the object now being protected. Move in close to the protected object and the meter reading should drop to zero. With some practice setting of R4 the circuit can be set to detect an intruder several inches away or the sensitivity can be reduced to a point where the object must be touched to trigger the sensor.

When using electronic circuits like this proximity sensor, it's sometimes tempting to push the sensitivity to its maximum range, and under ideal conditions you might get away with it. But in many applications where this type of sensor is used to protect a valuable object, conditions do not always remain ideal. Large changes in the humidity, rodent, component changes, or a number of other unforeseen environmental variations are just some of the problems that can cause a too-sensitive sensor to false alarm.

The proximity sensor circuit in Fig. 3-16 is designed for the electronic enthusiast who enjoys experimenting with a project rather than just building a cut-and-dried kit type of circuit. If this description fits you, then give the IC prox-

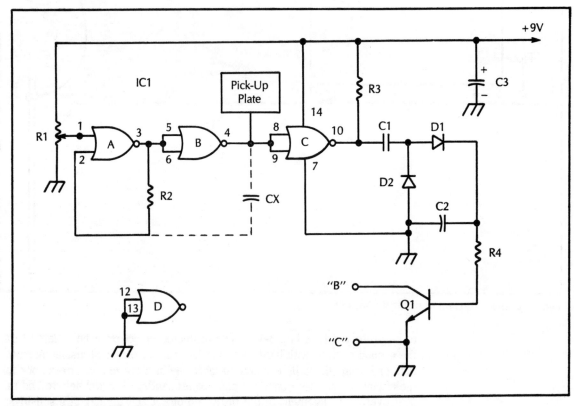

Fig. 3-16. IC proximity alarm sensor.

Table 3-6. Parts List for Fig. 3-16.

C1	.01 mF 100V disc capacitor
C2	.1 mF 100V disc capacitor
C3	100 mF 16V electrolytic capacitor
D1, D2	1N914 silicon diode
IC-1	4001 quad two-input nor gate
Q1	2N3903 npn silicon transistor
R1	50 kΩ pot
R2	2.2 MΩ ¼ W resistor
R3	1 kΩ ¼ W resistor
R4	3.3 kΩ ¼ W resistor
Cx	See text
Misc.	Cabinet, perfboard, wire, solder, etc.

imity sensor circuit a whirl. A word of caution is in order before starting the project. The 4001 quad two-input NOR gate is constructed with MOS P-channel and N-channel enhancement mode devices that contain circuitry to protect the gate's input against damage due to static voltages or high energy electric fields, but care should be taken when handling or applying power to the IC. The maximum voltage applied to any pin of the IC should not be greater than 16 volts dc and don't solder, add, or remove parts to the circuit while the power is on. Just take normal care when working with this or any circuit using CMOS devices and all should go well.

A 3×4-inch piece of perf board, push-in pins, and IC socket is a good way to breadboard the proximity circuit. Connect a 6×6-inch piece of p.c. board material, or a piece of metal, to pin 4 of the IC to serve as the sensor's pick-up plate. Keep the distance between the circuit and pick-up plate as short as possible to help reduce the possibility of picking up noise or 60 Hz hum.

Here's How the Circuit Works

IC gates A and B are connected to form a low frequency rf oscillator circuit with the operating frequency determined by the resistance value of R2 and the distributed capacitance around the IC, circuit components, and the IC's socket. Cx represents the distributed capacitance in the circuit diagram and the actual value is probably less than 25 pF. If the circuit fails to operate because its construction does not produce enough distributed capacitance around the IC, a 5 to 25 pF can be added between pin 2 and pin 4 of the IC, or a small trimmer capacitor can be used and adjusted for best operation of the circuit.

The oscillator's output level is controlled by R1, which sets the bias on the input of gate A, and serves as the sensitivity control for the sensor. R1 is set just to point where Q1 turns on giving a normal closed output at terminals B and C. Move in close to the pick-up plate and the oscillator's output should drop, or stop letting Q1 turn off and produce an alarm or open circuit condition at the output terminals.

Setup and operation of this IC sensor can follow the same procedure given for the proximity circuit in Fig. 3-13. Experiment with the sensor circuit all you want, but always observe the handling procedures needed for the CMOS IC.

The next proximity sensor circuit operates in a different manner than the first two circuits, and to my knowledge is an entirely different scheme than any other now in use. A low frequency transmitter and companion receiver are used together in such a manner that an intruder acts as a coupling agent between the two circuits causing an alarm output to be given. Each of the circuits are connected to its own antenna plate and located in a manner that offers a minimum of coupling between the two. When a person or large conductive object comes into close proximity to the transmitter's antenna plate that person or object acts as a larger radiating antenna to direct the rf energy to the receiver's antenna plate. When properly installed, this type of tx/rc proximity sensor is ideally suited to protect almost any valuable metal item.

The transmitter circuit is the simple two-transistor circuit shown in Fig. 3-17. Transistor Q1, L1, C1, and C2 together make up a low frequency Colpitts oscillator circuit with Q2 connected in an emitter follower stage that offers isolation between the oscillator and antenna plate.

The receiver circuit in Fig. 3-18 consists of a tuned input circuit (L1 and C1), a transistor Darlington amplifier that isolates the tuned circuit from any

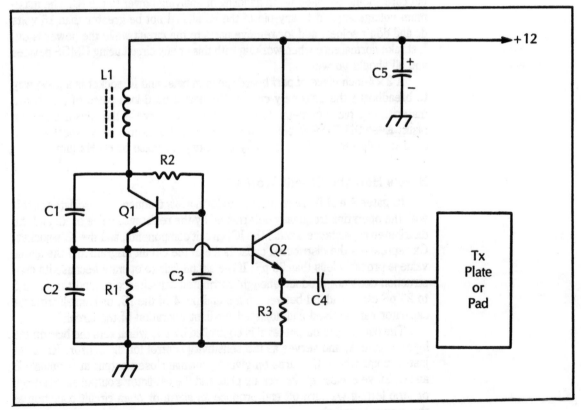

Fig. 3-17. Transmitter circuit.

Table 3-7. Parts List for Fig. 3-17.

C1, C2	.0072 mF 100V mylar capacitor (use two .0036 in parallel)
C3, C4	.1 mF 100V mylar capacitor
C5	100 mF 16V electrolytic capacitor
Q1, Q2	2N2222 npn silicon general purpose transistor
L1	75 turns of number 22 copper wire on a ferrite rod ⅜ inches × 3 inches long (see drawing in Fig. 3-14)
R1	1 kΩ ¼ W resistor
R2	220 kΩ ¼ W resistor
R3	2.2 kΩ ¼ W resistor
Misc.	Perfboard, push-in pins, wire solder, cabinet, transmitter plate, etc.

loading by the amplifier and detector circuitry (Q1 and Q2), and a normally open circuit transistor switch, Q4. The protected metal object is connected to the receiver's L/C tuned circuit, and functions as a pick-up antenna.

Building the T/R Alarm Sensor

Both circuits can be built on perf board and housed in a metal or plastic cabinet, and since neither circuit is critical, any good layout and wiring scheme should be okay. The two tuning inductors (L1 in the transmitter and L1 in the receiver) are wound just alike, with 75 turns of number 22 copper wire on a ⅜-inch by 3-inch ferrite rod as shown in Fig. 3-14. On the receiver's circuit board layout, leave some extra room in the area around C1 so the capacitor value can be fine tuned to bring the receiver's tuned circuit to the same exact frequency as the transmitter. If the protected object is very large, a coupling capacitor might be needed between the receiver's tuned circuit and the pick-up plate or protected object. If one is needed, try a .005 μF 100 volt mylar capacitor, as this size should work in almost all installations and does not cause any loss in the sensor's sensitivity.

Testing and Placing the T/R Sensor in Service

Select an object to be protected and connect the receiver circuit to the object as shown in Fig. 3-19. Locate a metal plate for the transmitter's antenna under a rug or cover on the floor, or mount it on an adjacent wall. The antenna plate should have a minimum surface area of 4 to 6 square feet, but a smaller size antenna can be used; it will give a narrower area of pick-up sensitivity. The only way to obtain the desired results in the circuit's sensitive and detection range is to experiment with different sizes and locations of the transmitter's antenna.

With all circuits connected to their antennas and power on to both, connect a dc (0 to 10 volt scale) voltmeter positive lead to the cathode of D1 and the negative lead to circuit common. Stand away from the transmitter's antenna plate and turn the receiver's gain (R6) all the way up, and if the voltmeter indi-

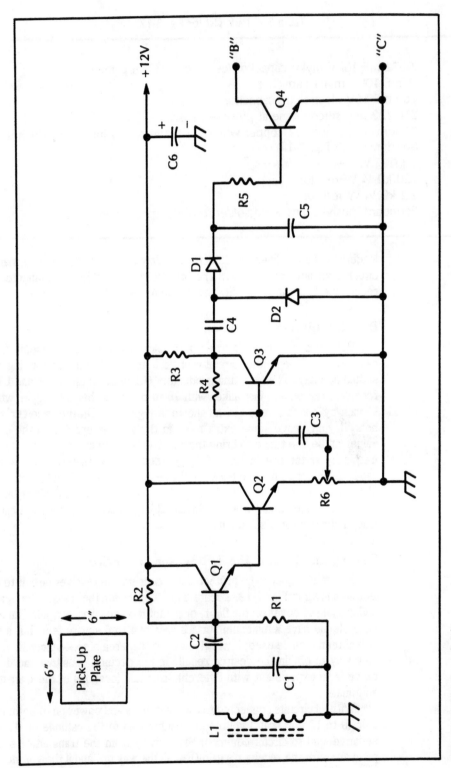

Fig. 3-18. Receiver circuit.

Table 3-8. Parts List for Fig. 3-18.

C1	.0036 mF 100V mylar capacitor
C2	39 pF 100V disc capacitor
C3	.01 mF 100V disc capacitor
C4	.068 mF 100V mylar capacitor
C5	.22 mF 16V mylar capacitor
C6	100 mF 16V electrolytic capacitor
D1, D2	1N914 silicon diode
L1	Same as L1 in Fig. 3-17
R1, R2	1 MΩ ¼ W resistor
R3	3.3 kΩ ¼ W resistor
R4	220 kΩ ¼ W resistor
R5	4.7 kΩ ¼ W resistor
R6	10 kΩ pot
Q1	2N5089 npn high gain transistor
Q2, Q3, Q4	2N3903 npn general purpose transistor
Misc.	Perfboard, pins, wire, pick-up plate, etc.

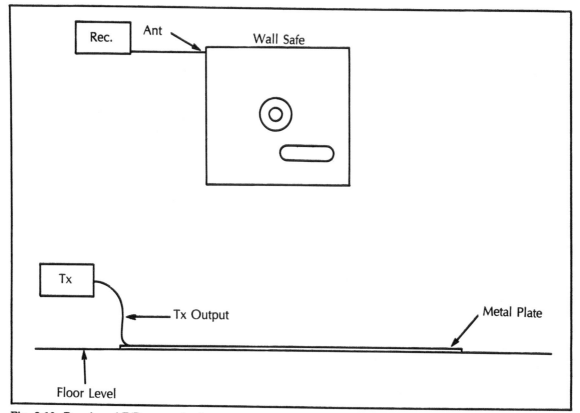

Fig. 3-19. Drawing of T/R sensor hook-up.

cates no more than .4 volts dc maximum, the gain can be set to its most sensitive setting. For best operation, the receiver's gain should be set to produce an output of .25 to .4 volts without anything in close proximity to either antenna.

An extra degree of care is in order when setting up the T/R sensor when it is used to protect a valuable object, as even the professional burglar probably has never faced this type of proximity sensor before. Do a good job in hiding all signs of the sensor installation and cover all traces of the leads going to and from each of the antennas.

Properly installed, even the simplest of the make and break alarm sensors can fool the best in the trade much better than a poorly-arranged high dollar super-sensitive sensor that can be detected, because, like most rats, the burglar is always keeping an eye out for the trap that could temporarily end his career of crime. Thoughtful planning as the first step before reaching for a screwdriver or electric drill goes a long way in ending up with a first class installation. So plan the overall scheme first, and then precede on with the actual installation with extreme care to detail and circuit wiring.

Ultrasonic sound waves can be used as the working medium for a special type of proximity alarm sensor that can be set to operate in either a normally closed or normally open sensor output condition. The drawing in Fig. 3-20 helps in understanding the operation of this simple, but useful, alarm sensor. The top drawing shows the transmitter's transducer aimed toward the receiver's pickup transducer, and as long as each is on and working, a N.C. sensor output will be present at the receiver. To catch a thief, he or she must move between the two transducers and cause the signal to stop reaching the receiver's transducer. When the signal path is broken, the N.C. output reverses to an open circuit sensor condition giving the master alarm control unit the go ahead to sound the alarm. Operating the tx/rx units in the continuous signal N.C. circuit condition offers a maximum of security, because anything that prevents the signal frame being present, or from reaching the receiver, automatically sends an alarm output signal to the master control unit. Also, if any of the circuit wiring is tampered with or if any of the wires are cut the alarm signal is given as if a true signal break has occurred.

The setup shown in the bottom drawing can be arranged to operate in either the N.O. or N.C. output condition. As shown, the transmitter is sending out a signal that's hitting a solid object and is reflected back toward the receiver's pick-up. If the signal is continuous and of sufficient strength the receiver, produces a N.C. circuit output, and any object that moves through the path causes an alarm output. But if the solid object is removed and the signal is no longer bouncing back to the receiver a N.O. circuit output is the sensor's normal condition. When an object of sufficient size and density moves close enough in front of the two transducers a signal is reflected to the receiver and a closed circuit output occurs, sending an alarm signal to the master control unit.

Either of the two methods of transmitting and receiving the ultrasonic signal produces a good and practical type of an alarm sensor. The actual end use helps in selecting the sensor that will operate best for the application at hand.

Refer to the basic sensor setup in Fig. 3-20 to help in determining the one best suited for the room or area that the sensor is to be used in.

A TWO-UNIT ULTRASONIC ALARM SENSOR

Figure 3-21 is the schematic diagram of the transmitter section of the sensor. A 555 IC is the working part of the circuit, with its operating frequency set by the values of R1, R4, and C1.

The ultrasonic transducer, TR1, is very much like a high Q tuned circuit that produces a greater output at its natural resonant frequency than at any other frequency, and must be driven by a signal source operating at this same frequency. If the oscillator driving the transducer drifts off frequency too far, the

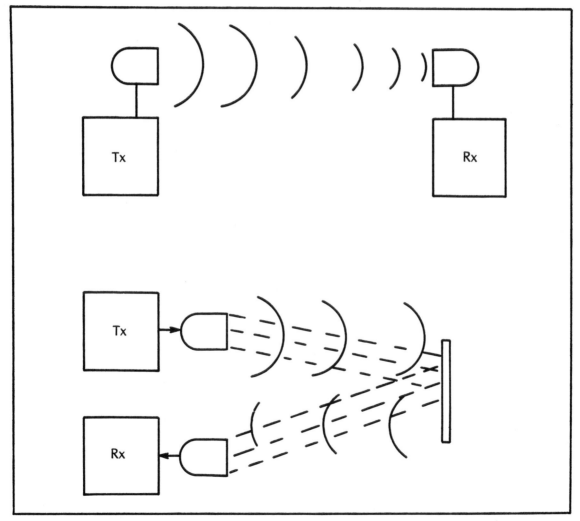

Fig. 3-20. Ultrasonic alarm sensor setup.

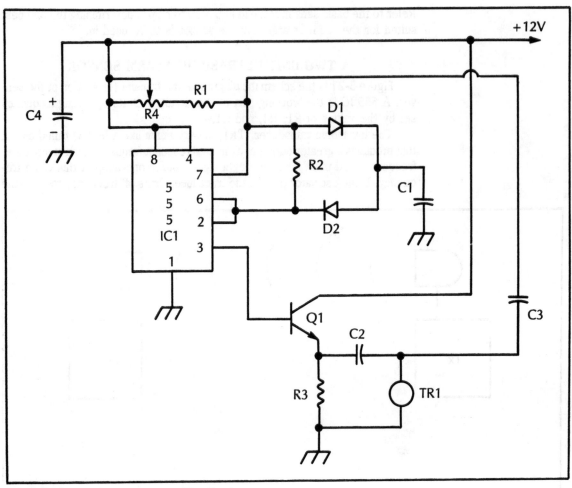

Fig. 3-21. Ultrasonic sensor transmitter circuit.

Table 3-9. Parts List for Fig. 3-21.

D1, D2	1N914 silicon signal diodes
C1	.011 mF 100V mylar capacitor
C2	.018 mF 100V mylar capacitor
C3	680 pF 100V disc capacitor
IC1	555 timer IC
Q1	2N3904 npn transistor
R1, R2	2.2 kΩ ¼ W resistor
R3	1 kΩ ¼ W resistor
R4	1 kΩ trim pot
TR1	40 kHz ultrasonic transducer
C4	100 mF 16V electrolytic capacitor
Misc.	8-pin mini-dip IC socket, perfboard and pins, cabinet, 12Vdc supply source, etc.

ultrasonic signal emitted from it drops in level and causes a false alarm signal to appear at the sensor's output. To add a degree of frequency stability to the R/C 555 oscillator circuit, a small amount of the output signal is sampled at the transducer and is coupled back to the IC through capacitor C3. Since the transducer is operating like a parallel resonant tuned circuit, the signal voltage present across it is greater at resonance than at any other frequency. This positive feedback tends to hold the 555 oscillator to the natural frequency of the transducer and offers a degree of frequency locking as the circuit is tuned with R4.

A regulated power source can add to the oscillator's circuit and frequency stability, but if a reasonably stable power source is used no problem should occur. A voltage variation of 1 volt or less should not bother the stability of the circuit's output frequency and signal level.

The transmitter circuit components can be mounted on perf board with push-in pins, and the completed circuit housed in a metal or plastic cabinet. The circuit is non-critical, and with normal care and a good layout scheme no trouble should occur in getting the transmitter to function properly. Since the number of component parts for the circuit is so few it would be a good idea to keep the transducer and circuit together in a single housing. Long runs of shielded leads will affect the operation and output of the transducer, and could cause the feedback circuit to not function properly. No matter what construction method you use take care to keep the lead length between the transducer and the oscillator circuitry as short as possible. Any length under 6 feet will do just fine.

After the wiring is completed and power applied, the next step is to set the oscillator's frequency to that of the transducer's natural resonance frequency. If an oscilloscope is available connect the input probe to the junction of C2 and C3, and set the vertical gain to 1 volt per cm.

As R4 is turned, very little happens until the oscillator hits the resonant frequency of the transducer. At that point of the pot's adjustment the oscillator's output increases considerably indicating that the resonant frequency of the transducer has been reached. This is the only adjustment necessary on the transmitter circuit and it can be put aside until the receiver is ready for testing.

Figure 3-22 shows the schematic diagram of the ultrasonic receiver circuit. Transistors Q1, Q2, and Q3 are all common emitter amplifier stages that raise the minute ultrasonic signal picked up by the transducer, TR1 to a level sufficient to be detected and to drive the sensor output transistor Q4 for either a normally closed or open circuit output condition. The receiver's overall gain is set with a 500-ohm pot (R13) that varies the ac emitter resistance of the final gain stage Q3. A voltage doubling rectifier circuit converts the ac ultrasonic information to dc that drives the base of Q4 that controls the sensor's output condition. The drawing in Fig. 3-20 should help you determine if a normally open or closed output is desired.

The receiver can be constructed using the same method used in building the transmitter circuit, or any good scheme can be used as layout and circuit wiring are not critical. Like the transmitter, it would be advisable to keep the leads connecting the transducer and receiver circuit as short as possible. Either a metal or plastic cabinet will do to house the completed circuit.

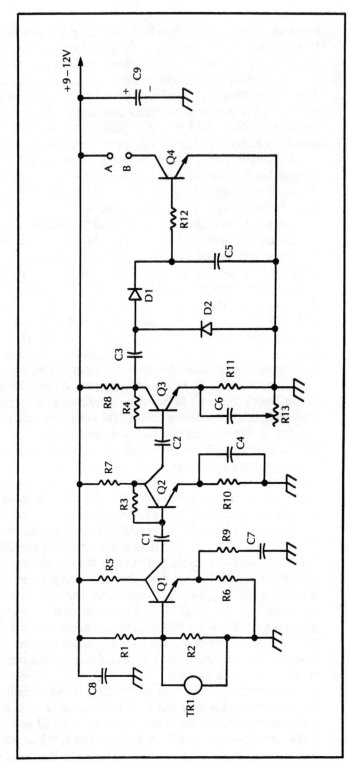

Fig. 3-22. Ultrasonic sensor receiver circuit.

Table 3-10. Parts List for Fig. 3-22.

C1, C2, C3	.02 mF 100V mylar capacitor
C4, C5, C6	.1 mF 100V mylar capacitor
C7	.068 mF 100V mylar capacitor
C8	.27 mF 100V mylar capacitor
C9	470 mF 25V electrolytic capacitor
D1, D2	1N914 silicon signal diode
Q1-Q4	2N3903 npn silicon transistor or similar GP type
R1, R2, R3, R4	220 kΩ ¼ W resistor
R5, R6, R7, R8	2.2 kΩ ¼ W resistor
R9	330 Ω ¼ W resistor
R10	470 Ω ¼ W resistor
R11	270 Ω ¼ W reistor
R12	4.7 kΩ ¼ W resistor
R13	500 Ω pot
TR1	40 kHz ultrasonic transducer
Misc.	Perfboard, pins, cabinet, 9 to 12Vdc power source, etc.

Placing the Two-Unit Ultrasonic Alarm Sensor in Service

If the set-up scheme shown in the top drawing of Fig. 3-20 is followed, the first test to make is to see just how far apart the two units can operate successfully. Choose an area without air currents. Locate the transmitter at a fixed position about three feet off of the floor and aim it down a long hall or open area. Connect a temporary power source to the receiver and set the gain control, R13, for a minimum resistance (maximum gain setting) and connect a dc voltmeter to terminals A and B to monitor the sensor's output condition. With a signal hitting the receiver's transducer, the meter reads a voltage near that of the power source.

Line the two transducers up one facing the other and move slowly away with the receiver unit until the meter reading begins to fluctuate or drop to near zero. Check out the last foot or two of the maximum operating range to see if the signal is actually strong enough for a reliable operating setup.

When installing any ultrasonic alarm system, there are several do's and don'ts to follow for a successful installation.

1. Don't feed a signal across an area where a heating or cooling system is moving air. To do so causes a false alarm almost every time the system cycles on and off.
2. Don't place any moving signs, hanging signs, or anything that can move when hit by an air current in the path of your ultrasonic signal.
3. Don't try to use an ultrasonic system outdoors, or indoors where a window or door is open.
4. If animals or birds run freely within the area you want protected, forget about using any ultrasonic system.

Another method in which the two-unit sensor can be used is to monitor a reflected ultrasonic signal from a hard surface such as a wall or door. If an object with a soft texture, like a person's clothes, moves between the hard surface and the two transducers, the receiver circuit responds with an alarm output. If a door is to be monitored, the sensor can be set up to sound off when the door is opened.

By placing the two transducers side-by-side, separated by no more than two inches and aimed in the same direction, this can function as a proximity detector to respond to a person or object that comes within a few inches of the two transducers.

No matter how you intend to use the two-unit sensor always keep one thing in mind: Don't push the range or the sensitivity to its maximum limit or install it in a hostile environment and you will probably end up with a first-rate detection system.

A SINGLE UNIT SENSOR

The circuit diagram in Fig. 3-23 is of an unusual single unit T/R ultrasonic alarm sensor.

A single IC generates the transmitter signal and at the same time is used as a selective receiver tuned to detect the reflected self generated signal. Sounds

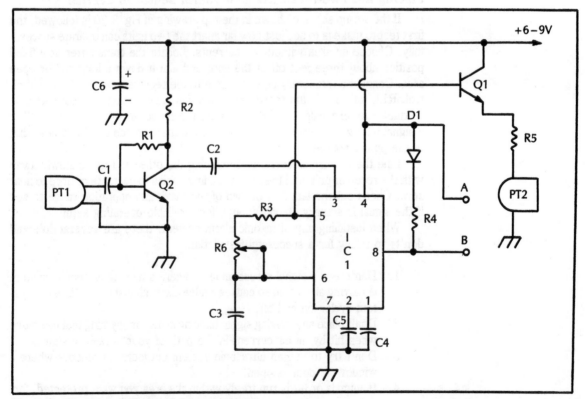

Fig. 3-23. Single unit T/R ultrasonic sensor circuit.

Table 3-11. Parts List for Fig. 3-23.

C1	.01 mF 100V disc capacitor
C2	.0036 mF 100V mylar capacitor
C3	.005 mF 100V mylar capacitor
C4	.47 mF 100V mylar capacitor
C5	6.8 mF 35V electrolytic capacitor
C6	220 mF 16V electrolytic capacitor
D1	LED; almost any will do
PT1	Piezo tweeter, Radio Shack type 2 or 3.5-inch size
PT2	2-inch CT-5 8 Ω voice coil tweeter; Calrad or similar type will work
Q1	ECG 152, SK 3054, or similar tab plastic case npn power transistor
IC1	567 PLL IC
Q2	2N3904 npn GP transistor
R1	220 kΩ ¼ W resistor
R2	2.2 kΩ ¼ W resistor
R3	4.7 kΩ ¼ W resistor
R4	1 kΩ ¼ W resistor
R5	47 Ω 2 W resistor
R6	25 kΩ pot
Misc.	Perfboard, pins, 8-pin IC socket, cabinet, 6–9Vdc power source, etc.

complicated? Not so, thanks to the versatile 567 PLL IC that performs as a signal source and detector all rolled up into one small 8-pin plastic chip.

Now let's take a close look inside the sensor circuit of Fig. 3-23 and see how the simple circuit performs these dual jobs. The ultrasonic information is picked up by a piezo tweeter that serves as the input transducer, and feeds the small signal to a single stage transistor amplifier where it is boosted to drive the input of the PLL IC, pin 3. Since the 567 is generating the received signal the IC is automatically tuned to detect its own frequency. No matter how much the frequency might drift away from a preset frequency, the receiver stays in step all the way. Using the transducers, there is no need to worry about the actual operating frequency, as was the case with the first two-unit sensor.

The operating frequency is set by the values of capacitor C3 and the series resistance of R3 and R6. The operating frequency can be tuned or changed by adjusting the value of the pot R6. With the component values given, the tuning range is approximately 8 to 25 kHz, and is limited only by the specifications of the two transducers used.

Pin 5 of the IC produces a square waveform signal at the tuned frequency and for isolation purposes is buffered by an emitter follower stage, Q1, and its output drives the 2-inch 8-ohm tweeter, PT2. This is all there is to the transmitter circuit, and as uncomplicated as it appears it does an excellent job supplying a strong source of ultrasonic energy for the sensor.

As long as the IC receives a signal of sufficient strength at its tuned frequency the LED D1 is turned on and a normally closed output is supplied at terminals A and B. When the ultrasonic signal path is blocked or interrupted, the sensor's output goes from a normally closed circuit condition to an alarm or open circuit output. The sensor can be used in either of the configurations illustrated in Fig. 3-20.

Actually, the sensor operates best when its operating frequency is subsonic rather than operating at a true ultrasonic frequency.

The best operating range for the sensor is dictated primarily by the frequency response of the two transducers used for PT1 and PT2, and with the two specified in the parts list the best operating range is from about 8 kHz to 16 kHz. If a much higher operating frequency is desired, select a pair of transducers rated for the desired operating range, as the circuit is designed to work with frequencies as high as 25 kHz, and by changing the values of C3, R3, and R6, the upper limit could go theoretically to the IC's specified limits. Of course, this would be too high for any practical use, as available transducers rated above 43 kHz are difficult to obtain.

With the two specified transducers, the circuit performs very well at an operating frequency of 12 kHz, and even though you can hear the transmitter's output coming from PT2, it is of little importance, because the likelihood of anyone being present when the sensor is on is unlikely. It is also possible, while operating at this frequency range, that a number of rodents might like it somewhere else better than at the sensor's location.

Building the Single Unit Ultrasonic Sensor

The circuit can be constructed breadboard style on perf board with push-in pins, and since the circuit uses so few parts any closely knit layout scheme should work just fine.

The two transducers do not need to be in close proximity to the main circuit, but if a long run is required use a shielded cable for each transducer. Don't try to save money by using a single shielded cable because the internal coupling between the input and output wiring will feed a false signal to the receiver and cause the sensor to operate poorly or not at all.

Testing and Placing the Sensor in Service

If the circuit component parts are connected properly and a power source of 6 to 9 volts is applied to the circuit, it is ready to detect its own output signal. Adjust the frequency control, R6, to approximately midposition and you should hear the high frequency tone. Now position the transmitter's transducer on a table or workbench and aim it towards an open area without any walls or large objects within 10 feet. With the LED in view, take the receiver's transducer and aim it at the output end of the transmitter's transducer and the LED should light. Walk while keeping the two transducers facing each other, and see how far you can separate the two units and still receive a useable signal. When the LED goes dark, the outer limit of the operating range has been reached.

A number of good locations are best suited for the ultrasonic sensor and the following are just a few that you can choose from.

1. Across a hallway
2. Across an entry
3. Across the front of a floor or wall safe, or valuable painting
4. Across an attic opening or basement entry
5. Just about anywhere a burglar might enter or travel through to reach a valuable item or a restricted area

Good judgment and common sense will help you decide where the sensor should be installed to do the best job in protecting the area.

VIBRATION SENSORS

A specialized and often overlooked alarm pick-up device is the vibration sensor. There are several reasons that this useful sensor is overlooked, because in the past many of these so-called vibration sensors were no more than a weighted mechanical contact arrangement that, after a period of time, caused more grief than any benefit they could provide. Almost any type of mechanical device that transfers current from one circuit to another in time will fail or at best give a hit and miss performance. This is one area in which electronics has proven its worth time after time by replacing a set of mechanical contacts with an electronic equivalent circuit; that is what makes the vibration sensor circuit in Fig. 3-24 work so well.

It's not so much the electronic circuit that makes this sensor a success, but the unusual pick-up that detects the vibration. Take a look at Fig. 3-25 and you will see a mini loudspeaker modified to be very sensitive to low frequency vibrations and at the same time ignore any sound that is transferred through air to the speaker's cone.

A single IC containing four separate op amps (LM324) takes care of the electronic functions for the sensor. The first amplifier A is connected as a source follower that matches the low impedance of the speaker's voice coil, while the second and third amp B and C are gain stages that bring the level up 2000 times. The gain can be made even higher by changing R6 to a 1 k resistor raising the gain to about 10,000, but for most applications the parts values given will do fine.

The amplified signal is passed through C5 to a voltage doubler circuit, and the positive dc output is fed through a current-limiting resistor R8 to the base of transistor Q1. While the circuit is on and receiving no output from the pick-up, the dc voltage at the base of Q1 is normally zero, leaving it in the off condition, allowing Q2 to receive a forward bias through R10 and R11 turning it on to produce a normally closed output at terminals A and B. When a signal is picked up and amplified, the dc voltage at the output of the doubler drives Q1 on and removes the forward bias going to Q2, causing it to turn off, giving an open circuit at output terminals A and B.

If, for some reason, the application for the vibration sensor requires an output condition of a normally open circuit, a simple modification to the circuit can offer the output reversal. Remove Q2, R10, and R11, make the collector of Q1 the new output terminal B, and A terminal remains unchanged. Any of the sensors so far described is set up for a solid-state output, but in most cases each circuit

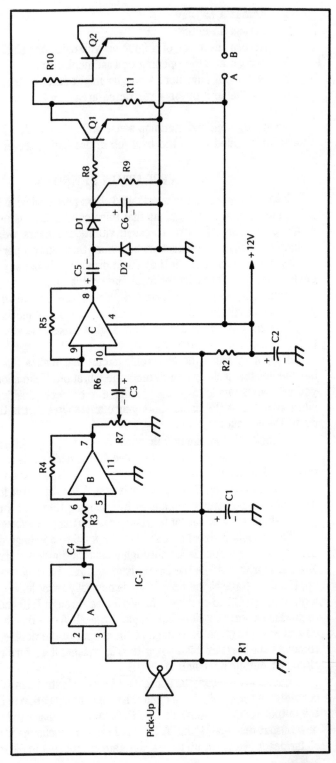

Fig. 3-24. Vibration sensor circuit.

Table 3-12. Parts List for Fig. 3-24.

C1, C2	100 mF 16V electrolytic capacitor
C3, C5, C6	6.8 mF 25V electrolytic capacitor
C4	.1 mF 100V mylar capacitor
D1, D2	1N914 silicon signal diode
IC-1	LM324 quad op amp
Pick-up	1½-inch 8 Ω speaker with modification (see text)
Q1, Q2	2N3904 npn GP transistor
R1, R2, R3	1 kΩ ¼ W resistor
R4, R5, R9	100 kΩ ¼ W resistor
R6	4.7 kΩ ¼ W resistor
R7	5 k Ω pot
R8, R10, R11	10 kΩ ¼ W resistor
Misc.	Perfboard, pins, penny, sewing needle, etc.

can have a sensitive relay connected to the A and B terminals, and offer a dry set or sets of contacts to connect to any master control unit. More information about how to select and connect these and other sensors to a master unit is given in a later chapter.

Building the Vibration Sensor

The first item to make up is the sensor's pick-up, and Fig. 3-25 should help in duplicating one of your own. Almost any or 16-ohm small speaker with a cone size of 2 inches or less can be used for making the pick-up. The 1.5-inch unit was selected because a copper penny covers the voice coil area and a good portion of the speaker's cone. This amount of coverage helps the sensor ignore any outside sounds that could give the amplifier a false signal and cause an alarm signal to be sent out.

Take a shiny new penny and solder a sewing needle to the center, so it protrudes out perpendicularly from the face of the coin (see the side view in Fig. 3-25). This arrangement is useful when a window, safe, or other item that's to be protected can have the sensor's pick-up placed where the needle can sit softly on a surface critical to movement on the protected item. It's easy to see, with the sensor's needle against a plate glass window, that when it is hit or broken a definite vibration will occur and be detected with the sensitive pick-up.

If a weighted pick-up is desired, the needle can be replaced with a short piece of stiff wire soldered to the penny's center and bent down and parallel with the speaker cone. Attach a small weight to the end of the wire (another penny will do) and you now have a pick-up that is sensitive to vibrations that includes the pick-up unit as well as the item that is supporting it. A good example of this type of installation would be to hang the pick-up in a position so the weighted wire is suspended vertically, making the unit sensitive to vibrations from any of the four sides of the pick-up and enclosure.

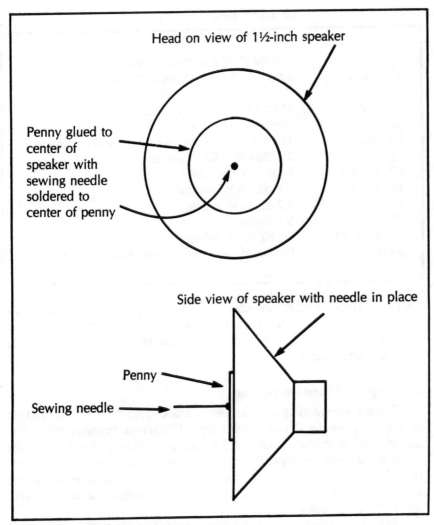

Head on view of 1½-inch speaker

Penny glued to center of speaker with sewing needle soldered to center of penny

Side view of speaker with needle in place

Penny

Sewing needle

Fig. 3-25. Vibration pick-up drawing.

The electronics can be built breadboard style on perf board, or a circuit board layout can be made. In any case the circuit is non-critical and if a good construction form is followed no trouble should occur in getting the circuit to work on the first try. Either a metal or plastic cabinet will do to house the circuitry, or the circuit can stand alone without any enclosure. Use whatever works out best for you and the installation.

Testing the Vibration Sensor

Connect power to the circuit and set the gain control R7 to mid-position. Temporarily connect a 3.3 k resistor across the output terminals A and B. Hook the test leads of a dc voltmeter (dc range 15 or 20 volts full scale) across the 3.3 k resistor, positive lead to terminal A and negative to B. Without an input to the pick-up the meter should read close to 12 volts. If not, recheck the circuit wiring against the circuit diagram. A quick check to determine that each of the op amps is connected and working properly is to take the voltmeter and check the voltage between pin 1 and ground, pin 7 and ground, and pin 8 and ground. If all is okay, each should read about 6 volts, or half of the supply voltage used to power the circuit.

If all checks out take the pick-up, and (if you built yours like the one in the drawing) carefully position the speaker on about a 45° angle letting the needle rest softly on a workbench or table top. While watching the meter, tap on the table near the needle. The meter's reading should drop momentarily to zero, and after the tapping ceases, should rise back to its original reading. The circuit's gain can be set to pick up almost any vibration that would indicate a break-in. To reduce the possibility of a false alarm, never set the sensor's gain higher than is necessary to produce a good and reliable output for the application or installation.

The pick-up using the needle can be used several ways to protect any item that would require breaking or banging on to enter or reach the goods. The simplest application is to position the pick-up so the needle lies softly on the surface of the protected item. Always run several tests with all setups you use to ensure that the position and location of the pick-up detects any attempt to gain access to or cause damage to the protected object. Any alarm sensor is of little value if it is installed improperly or in a hurried manner without a complete check-out to see that it is indeed doing what's expected of it.

THE SECURITY FENCE SENSOR

Security fencing can be protected electronically by using a special type of vibration or movement detector to give out an alarm signal when the fence is moved or shook from an outside force. The previous circuit could be modified to work in this area, but the circuit in Fig. 3-26 is designed especially for this application and when properly installed does an excellent job. The major difference in this sensor is the pick-up, which is designed to be very sensitive to a back-and-forth movement. Also, a fairly good output is obtained if the sensor is exposed to an up-and-down movement, so no matter what someone is trying to do to enter a secure area through or over a security fence, the sensor can pick up the movement and send out an alarm. The other pick-up could be operated in any position, but the one in Fig. 3-27 must always be positioned vertically for maximum sensitivity.

Operation of the circuit in Fig. 3-26 is very simple and straightforward. Op amplifier A is connected in a source follower circuit for impedance matching with its output connected to the sensor's gain pot, R9. The vibration signal is fed from the wiper of the pot to a single high gain amplifier stage A of the IC.

The output signal is rectified by the voltage doubler and the dc output supplies a forward bias for transistor Q1. Without an input signal, the dc output of the doubler is zero, and Q1 is in a non-conducting state, allowing Q2 to receive a forward bias through R7 and R8, to produce a normally closed circuit at sensor's output terminals. When a vibration is detected, the dc voltage from

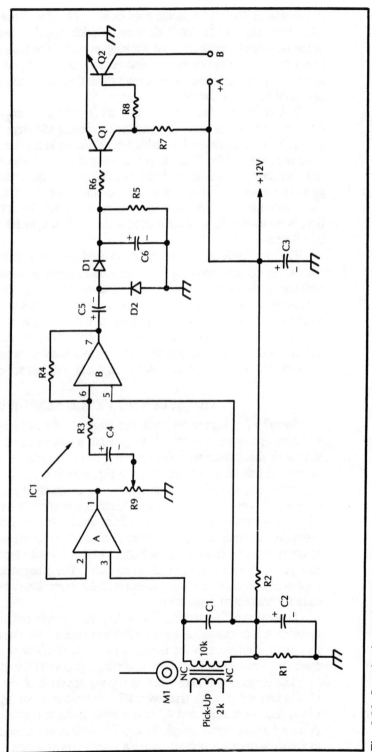

Fig. 3-26. Security fence alarm sensor circuit.

Table 3-13. Parts List for Fig. 3-26.

C1	.068 mF 100V mylar capacitor
C2, C3	220 mF 25V electrolytic capacitor
C4, C5, C6	4.7 mF 25V electrolytic capacitor
D1, D2	1N914 silicon signal diode
IC1	LM324 quad op amp IC
Q1, Q2	2N3904 npn transistors
R1, R2, R3	1 kΩ ¼ W resistor
R4	1 MΩ ¼ W resistor
R5	100 kΩ ¼ W resistor
R6, R7, R8	10 kΩ ¼ W resistor
Pick-up	10 kΩ to 2 k Ω mini audio transformer and misc. (see text)
M1	Magnet (see text)
Misc.	String, perfboard, two enclosures, wire, etc.

the doubler drives Q1 on and removes the bias from Q2, turning it off and sending out an alarm signal.

Building and Using the Security Fence Sensor

Here again is an electronic circuit that's non-critical in its circuitry layout; any good construction scheme will do. The only part of the sensor that should adhere to the suggested construction layout is the pick-up element shown in Fig. 3-27. Although there are several ways that the pick-up can be built once the principal of operation is understood, the one shown in the drawing has been built, tested, and performs as stated.

The best place to start in building the pick-up is to modify a small 10 k to 2 k audio transformer by removing all of the transformer's laminations.

Take all of the E-shaped pieces of the lamination and re-insert all, if possible, back in place in the transformer's core, all in the same direction, so the combined laminations look like a fat E. Trash the small I pieces of lamination material as they are not needed. All of the laminations may not fit back in place, so if one or two won't go that's okay as long as the total number used fit nice and snug in place. The 10 k winding of the transformer is used for the signal output and must be connected to the circuit through a section of shielded cable.

A plastic cabinet 6×1.5×1.5 inches or larger can be used to house the complete pick-up as shown in Fig. 3-27. The one item that's important to the construction of the pick-up is the arrangement of the donut magnet and the modified transformer. The magnet is suspended above with a piece of plastic or nylon string large enough not to stretch after a period of time and let the magnet reach the transformer's core. The spacing between the magnet and core should be between ³⁄₁₆ and ¼ inches. With this arrangement of the magnet and core, a definite attraction between the two should be obvious, as if an invisible spring connected the two together.

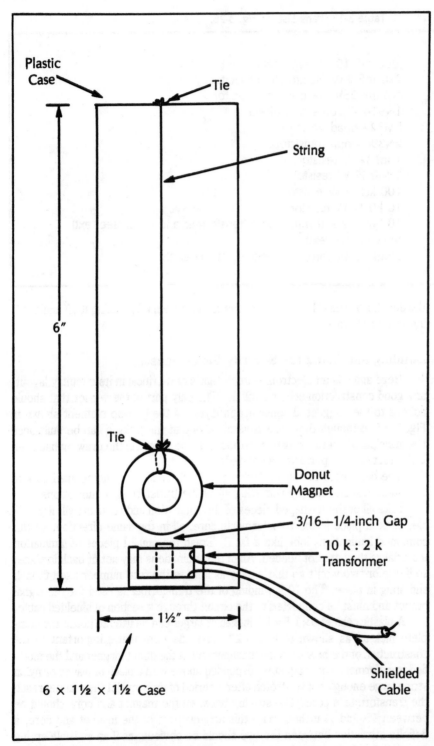

Fig. 3-27. Fence pick-up sensor drawing.

Labels in figure:
- Plastic Case
- Tie
- String
- 6″
- Tie
- Donut Magnet
- 3/16—1/4-inch Gap
- 10 k : 2 k Transformer
- 1½″
- 6 × 1½ × 1½ Case
- Shielded Cable

Two or even three of the input pick-ups can be connected in parallel to let only a single circuit function for up to three separate sensor locations along a length of security fence. The circuit gain needs to be set higher as the number of pick-ups are added, but the savings in circuit duplications far outweigh any inconvenience in making the necessary gain adjustments.

When making the final installation of the pick-ups, it is very important to have each of the units mounted on the fence in a near-level vertical position so the magnet can be free to swing in any direction equally before hitting the cabinet's inside walls. If the pick-up is mounted off center by only a small amount, you probably will never notice any difference in its sensitivity to movement from all sides, but if it is mounted too far off center it requires a greater amount of movement from any direction because the output is reduced any time the magnet starts its movement off from the center position over the transformers core. Beyond these simple installation precautions you should have no major problems in obtaining excellent results in using this unusual security fence sensor.

VEHICLE SENSOR CIRCUIT

The passing of a motor vehicle over a road or path into a secured area could mean trouble if it is not detected and a proper security check made. Unless a TV camera is used to keep an eye on the road and a security guard to keep an ever-vigilant eye on the monitor there is some instant when a vehicle might pass through unnoticed. The electronic sensor can do about the same job without keeping a person tied up, and as long as power is on it won't fall asleep and let someone enter unannounced. No, electronics is not always the only answer, or even the best answer, but when the cost of keeping an area secure is a budgeted item, then every practical avenue must be looked at before the best possible security system can be planned, no matter what the cost. The electronic vehicle sensor can even have its place in the most sophisticated of security systems, if for no other reason than to give an electronic wake-up call to an overworked security guard.

One electronic answer is shown in the circuit diagram in Fig. 3-28. The heart of this vehicle sensor circuit is designed around the Maxwell inductance bridge. The basic Maxwell bridge compares an unknown inductance with a known capacitor and their differences can be balanced out with a pair of variable resistors.

Usually the bridge circuit is driven with an ac signal source that can be detected and measured in level to indicate when the bridge is in or out of balance. If the ac output of the bridge circuit is amplified many times, then the circuit can detect minute changes in the inductor used in the bridge, and that's exactly how our circuit detects the presents of a vehicle by noting these small inductance changes caused by the vehicle passing over it.

Circuit Description and Operation

Now that you know how a Maxwell inductance bridge functions, let's see how the remaining circuit operates to detect a passing or parked vehicle. Transistor Q1 and its associated parts make up a Colpitts oscillator that's operating

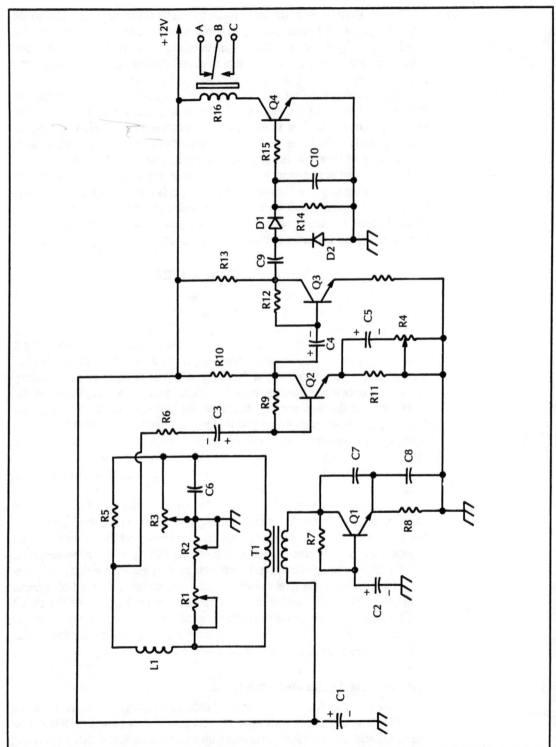

Fig. 3-28. Motor vehicle pick-up sensor circuit.

Table 3-14. Parts List for Fig. 3-28.

C1	470 mF 25V electrolytic capacitor
C2	4.7 mF 25V electrolytic capacitor
C3, C4	.47 mF 25V electrolytic capacitor
C5	47 mF 25V electrolytic capacitor
C6, C7	.27 mF 100V mylar capacitor
C8	.5 mF 100V mylar capacitor
C9, C10	.1 mF 100V mylar capacitor
D1, D2	1N914 silicon signal diode
R1	2 kΩ pot
R2	100 Ω pot
R3	2 kΩ pot
R4	1 kΩ pot
R5	100 Ω ¼ W resistor
R6	1 kΩ ¼ W resistor
R7	220 kΩ ¼ W resistor
R8	270 Ω ¼ W resistor
R9	680 kΩ ¼ W resistor
R10	4.7 kΩ ¼ W resistor
R11, R13	1 kΩ ¼ W resistor
R12	100 kΩ ¼ W resistor
R14	10 kΩ ¼ W resistor
R15	2.2 kΩ ¼ W resistor
Q1-Q4	2N3904 npn GP transistor
R16	1 k sensitive relay
L1	50-turn coil of #26 wire on 50-inch form
T1	600 Ω to 600 Ω coupling mini audio transformer
Misc.	Cabinet, wire, perfboard, shielded cable, etc.

at a frequency of about 10 kHz. The output is coupled to the bridge circuit through the secondary of T1. The loop inductor, L1, is balanced with pots R1, R2, and R3 to produce a near-zero level ac signal at the output of the bridge circuit. The bridge output signal is taken off at the junction of R5 and L1 and is fed to the base of a common emitter amplifier stage Q1. The circuit's gain is made adjustable with pot R4. The amplified signal is fed to the base of Q2, another common emitter amplifier stage, and its output drives a voltage doubler circuit.

The dc output of the doubler circuit supplies forward bias for the relay driver transistor Q4.

When the circuit has been properly balanced, the relay is in a normal non-operated state, and the relay contacts can be connected for either a normally closed or normally open sensor output condition. (Use contacts A and B for a normally closed circuit, or contacts B and C for a normally open circuit.) Assuming the large loop inductor L1 is undercover of a roadway, and a vehicle passes over the area where the loop is hidden, a change in the inductance of the loop occurs as the large metal object passes over it. The bridge is unbalanced and the relay operates, sending out an alarm signal. After the vehicle has passed the loop, the circuit again is in balance and the relay returns to its non-operated state.

Building the Vehicle Sensor

The component parts can be assembled on a piece of perf board with push-in pins and housed in a metal or plastic cabinet. Any good workable construction scheme will do, as the circuit layout is not critical, but if all of the adjustment pots are placed in line on one side of the cabinet, setting up the sensor is much easier.

Winding the Pick-Up Loop

The pick-up loop is made by winding 50 turns of number 26 enamel-covered copper wire around a 25-inch wood or plastic form. If a winding form with these dimensions cannot be located a slightly different diameter can be used by winding about 350 feet of wire on it. The winding can be made in a jumble-wound fashion and if a temporary coil form is used, the windings can be removed and kept in place with weatherproofing tape. In any case if your loop is like the one used in the circuit, it should balance okay in the bridge circuit with the component values given in the parts list. If the loop is located over 10 feet from the circuit, a shielded cable should be used to connect the two together.

Placing the Sensor in Service

The most important thing to consider when installing the sensor is the location of the pick-up loop. If the loop is to be buried underground it should be protected with weatherproofing tape or silicon rubber to keep moisture out. The loop should not be buried any deeper than is necessary to protect it from damage from traffic, and should be parallel to the surface of the roadway. Usually burying the loop two or three inches down offers sufficient protection.

The circuit and pick-up loop should be checked out together before the final burial of the loop. This is best accomplished by connecting the loop to the circuit through a length of shielded cable that will be used in the final installation. Locate the pick-up loop away from any large metal object and connect a power source of 12 volts to the circuit. Set each of the pots to mid position and observe the condition of the relay, as it is unlikely that the circuit is in balance so the relay will be operated. Connect the positive lead of a dc voltmeter to the cathode of D1 and the negative lead to circuit common, or ground. Set the voltmeter to read 5 or 10 volts full scale, and rotate R1 in a direction that reduces the meter's reading to its lowest reading. Repeat the same adjustment with R3 for the lowest reading. Now go to the fine adjust pot R2 and rotate for the sharpest null that you can obtain. Usually, to obtain the best circuit balance you need to go back and forth among the three pots.

The circuit gain control pot can be used to raise or lower the gain for proper operation of the relay. If the relay happens to pull in, even with the circuit in perfect balance, the gain control can be set slightly below the relay's drop out point for proper circuit operation.

After the pick-up loop is buried in place, the bridge circuit needs to be rebalanced to compensate for the ground effect on the inductance of the loop. The setting of the gain is very important and should not be set too high. The actual final setting should be determined by experimenting with different types of motor vehicles passing over the loop and a number of different gain settings. The proper combination that gives the most reliable operation with the lowest gain setting is the one to use.

It is also possible that resetting the balance and gain controls may be necessary from winter to summer to compensate for any temperature effect on the inductance of the loop pick-up coil.

Any number of vehicle alarm sensors can be placed at different locations along a roadway or any area where a vehicle might pass, but don't locate any of the loops closer together than fifteen feet or an interaction between the circuits makes balancing the bridge circuits impossible.

SOUND-ACTIVATED ALARM SENSORS

This method of detecting a burglar can be a most trying experience when trying to do it with electronic circuitry, but with proper care in the installation and application a successful sensor can be designed to fulfill a special need. Outside sounds, like car horns, thunder, and aircraft can give an electronic sound activated alarm sensor a real fit—a problem that's not easy to overcome.

The block diagram in Fig. 3-29 illustrates one method of circuit design that can offer a degree of success in detecting a break-in or burglar entering an area

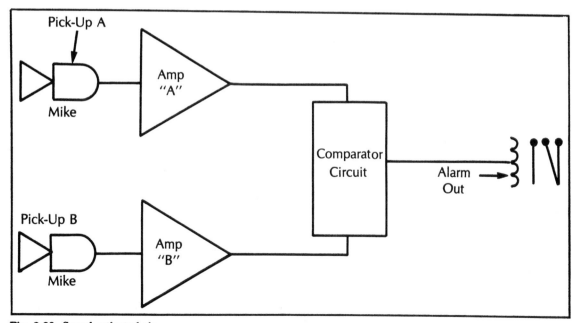

Fig. 3-29. Sound-activated alarm sensor.

and making a noise in the process. Two identical audio amplifier circuits with similar pick-up mikes located at two different locations in the same room or building can be used to eliminate most single-source outside noises like car horns or thunder.

The output of each amplifier is converted to a dc voltage that feeds a comparator circuit that looks at each dc level and their timing. If both levels are about the same and both arrive together, the output is ignored as a sound source coming from the outside and not an alarm condition.

If a noise is made near one mike and away from the other, the comparator determines that the sound is coming from a specific area and gives out an alarm signal. Although the theory seems sound, the actual operation doesn't always go that way. Of all alarm sensors, the sound-activated type gives out more false or undetermined alarms than many of the others.

So why even consider this type of an alarm sensor at all? In some locations this type of sensor can be most useful in detecting someone making loud noises and doing damage rather than actually trying to steal anything. Or a different type of sound sensor can be used to monitor an area with an audio amplifier connected to a threshold circuit that only comes on line when the sound level reaches a preset amount. This type of alarm can be useful in monitoring an area where a low sound level is a normal condition. No one wants to listen to the whine of an electric motor or machine all of the time, but a system that is only activated with a loud sound and stays keyed on until a manual reset is pressed is a very useful type of an alarm sensor, because it leaves the decision to a human mind to determine if an alarm signal should be sent out.

Due to the problems associated with the stand-alone sound sensor and the large number of false alarms it can produce, the threshold keyed system of utilizing sound as a means of detecting the presence of a burglar is the best way to proceed with this type of alarm sensor. If several of these sensors are used covering a number of rooms or buildings a one-person security operation could monitor a number of video monitors and not be bothered with the audio until one or more are keyed on.

How the Threshold Circuit Operates

The circuit diagram in Fig. 3-30 is of an audio alarm sensor that uses the threshold keyed system to offer a silent output until the amplifier is keyed on. A single quad op amp IC takes care of the circuit's gain needs, with amp A serving as a mike pre-amp, followed by a second gain stage B. The audio gain is controlled by a pot in the output of amplifier A. The amplified sound signal is connected to a source follower, amp C, which offers zero voltage gain, but isolates the gain stages from the voltage doubler circuit, threshold detector, and the keyed amplifier.

The normal sound picked up by the mike is amplified to a level great enough to operate the voltage doubler circuit which changes the ac audio to a varying dc voltage that follows the general waveform of the input. As long as the peak level of the dc varying voltage is less than the reference voltage set with the trigger adjustment pot R11 at the emitter of Q1, nothing happens. When a loud

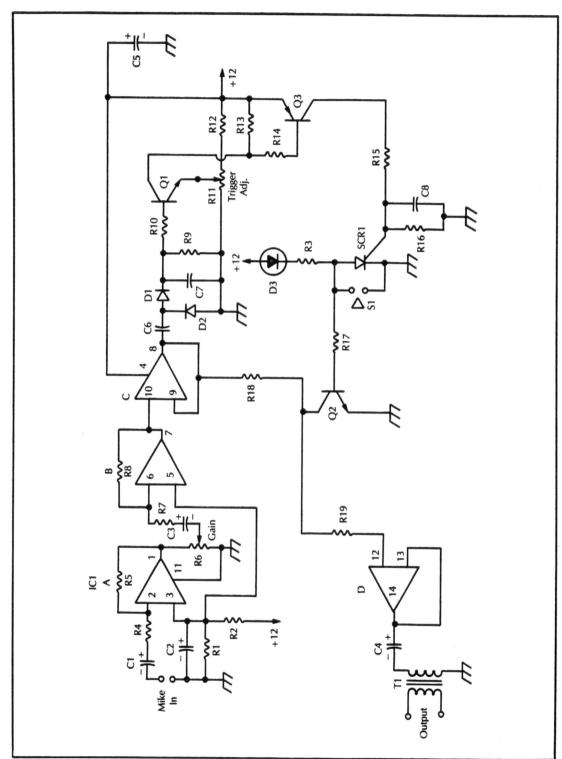

Fig. 3-30. Audio threshold alarm sensor circuit.

85

Table 3-15. Parts List for Fig. 3-30.

C1, C3	.47 mF 25V electrolytic capacitor
C2, C5	100 mF 25V electrolytic capacitor
C4	47 mF 16V electrolytic capacitor
C6, C7, C8	.1 mF 100V mylar capacitor
D1, D2	1N914 silicon signal diode
D3	LED (any color)
IC1	LM324 quad op amp
Q1, Q2	2N3904 npn silicon transistor
Q3	2N2906 pnp silicon transistor
SCR1	2N5062 or similar .8 A SCR
S1	N.O. push-button switch
T1	600 Ω to 600 Ω matching audio transformer
R1, R2, R3, R4	1 kΩ ½ W resistor
R5, R8, R9	100 kΩ ¼ W resistor
R7, R10, R13, R14, R17	10 kΩ ¼ W resistor
R6	10 kΩ pot
R11	1 kΩ pot
R12, R15, R18, R19	2.2 kΩ ¼ W resistor
R16	4.7 kΩ ¼ W resistor
Misc.	Low impedance mike, mike jack, perfboard, cabinet, wire, solder, etc.

noise of sufficient level is passed to the doubler circuit, the dc positive output turns Q1 on, supplying a forward bias for Q3, and turning it on also. The collector current of Q3 through R15 supplies a positive gate current for the SCR, triggering it on. As long as the SCR is off, the current through D3 and R3 keeps Q2 turned on, which acts as a switch that ties the junction of R18 and R19 to ground, keeping the audio signal from passing on to the source follower, amplifier D, that feeds the audio to the monitoring station through a matching transformer, T1. When the SCR is turned on, the bias is removed from Q2 and it switches to an open circuit, allowing the audio to pass on through. The audio channel remains open until the push button switch S1 is manually reset. The LED gives a visual indication of the circuit's condition.

Building the Audio Threshold Sensor

A handy way to build the circuit is to mount all of the components on perf board with push-in pins. A 14-pin IC socket should be used for the LM324, and a metal cabinet offers a degree of shielding from outside rf and high level 60 Hz signals that could interfere with the circuit's operation. The circuit layout is not critical, but it is a good idea to keep the components' leads short and direct to the point around the two gain stages of IC1. Both of the pots should be mounted on the cabinet in a convenient location for easy access.

If a long run is necessary between the circuit and the pick-up mike, shielded mike cable should be used. The input amplifier is designed to operate best with a low output impedance mike, but since audio quality is not the most important factor, try whatever mike you may have on hand and test it out to see how well it operates before buying a mike especially for the sensor.

The audio output is fed through a mini 600 Ω to 600 Ω audio coupling transformer, and is of sufficient level to drive almost any utility amplifier that would be used as a monitor.

Placing the Audio Sensor in Service

There's one important area that should be given some extra consideration, and that's the location of the pick-up mikes. The sensor stands or fails on where the mikes are located. Sometimes just a slight change or movement of a mike can make all the difference in how successful the desired area is covered. That means the only way to really determine if the best mike location has been found is to run a number of tests with the system in actual use.

After a tentative mike location has been selected and the shielded cable connected between the two, and with the output of the sensor connected to the input of the monitor amplifier, set all of the sensor's controls to mid-position and turn the power on. Since the unit is to be tested for the best operating performance, set the trigger adjustment pot R11 to the most sensitive position (wiper at the ground end of rotation). The first sound picked up by the mike will most likely trigger the SCR on and the audio will automatically pass through to the monitor amplifier. With the amplifier keyed on, setting up the system is much easier than trying to have the circuit trigger every time a single noise is made.

The sensor's gain can best be set with its output feeding the monitor amplifier, and with someone in the area where the mike is located to make sounds that might be heard in a breaking in. Adjust the amplifier's gain so the normal background noise produces a low-level output from the monitor amp. Now have the person near the mike make a noise like someone breaking in. If the mike is located in a good place, the sound level should be much louder than the normal background level. If the difference is very apparent, turn the threshold control for its minimum sensitive position and reset the amplifier with S1. Have the noisemaker repeat the sound at the same level at intervals of 20 seconds until you have had time to set the threshold pot to a level that keys on each and every time the sound is heard.

No matter how many different sensors you build to use with your burglar alarm, there must be some logical way in which to connect all of these sensors together to produce a reliable and practical alarm system. The burglar alarm control units covered in the next chapter offer several different circuits to choose from.

Chapter 4
Burglar Alarm
Control Systems

Now that we have a number of different alarm sensors to use for detecting a breach in security, the next item needed is a method of connecting all of these sensor pick-ups to a common control system. The alarm control systems described in this chapter are designed to operate with the sensors in Chapter 3 and the alarm indicators and sounders in Chapter 5 to make a complete burglar alarm system.

Since the majority of the sensors previously described have control outputs that are solid-state in nature and not hard contacts like relays, a means of connecting these to any universal control system is necessary, and the following interface circuits help fill that requirement.

If the output of the sensor circuit is of the type shown in Fig. 4-1 it can be connected to an interface circuit like the one in Figs. 4-2 and 4-3. The simplest interface circuit, as in Fig. 4-2, is a relay connected in the collect circuit of the sensor's output and the contacts used to transfer the information to the control system. The circuit in Fig. 4-3 offers total circuit isolation between the sensor and the control unit. The sensor's output drives an optocoupler by supplying current to the internal LED, which in turn activates the photo-transistor to either operate a relay or to drive a solid-state input circuit on the control unit.

These examples of connecting a sensor to a control system are not the only methods that can be used, but they give you an idea of what's needed for a typical alarm system. As the control system circuits evolve in this chapter a number of practical interfaces are shown.

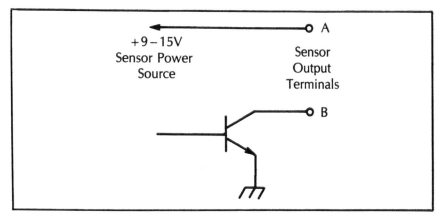

Fig. 4-1. Sensor output circuit.

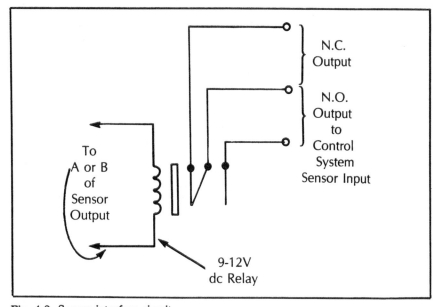

Fig. 4-2. Sensor interface circuit.

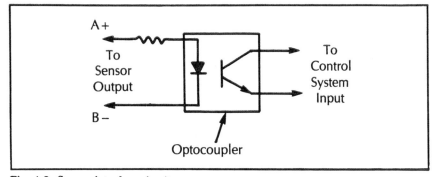

Fig. 4-3. Sensor interface circuit.

A BASIC RELAY-OPERATED CONTROL SYSTEM

The circuit in Fig. 4-4 is a bare-bones workable relay-operated alarm control system that uses normally closed input sensors and is similar to a circuit used professionally for many years throughout the world.

A sensitive low operating current relay is powered by a battery and remains in the operated condition as long as the sensor input terminals A and B are tied together electrically through the external sensors, and the batteries are kept fresh. In the earlier days, when this basic type of circuit was the mainstay, it used large telephone cells for power and a self-interrupting outside bell as the system's alarm sounder. Some installations used a separate set of batteries for the bell, but the majority used the same power source for both.

The types of sensors available at that time were mostly the simple mechanical break circuit style and only a very limited variety was available at best. A method used for a long time to indicate the breakage of glass windows and doors was lead tape routed around the outer edge of the protected glass area. Simple mechanically operated normally closed switches were used to indicate if a door or a window was in a secure position. Actual break wire systems existed that crossed over vertical openings and air passages. The best security of that day (and even with today's high tech equipment) is in the way that the sensors are set up to trap an intruder.

AN N.C. ALARM CONTROL SYSTEM

The alarm control circuit in Fig. 4-5 is well suited for a small to medium size installation, primarily where mechanical contact devices are used for the sensor output circuits.

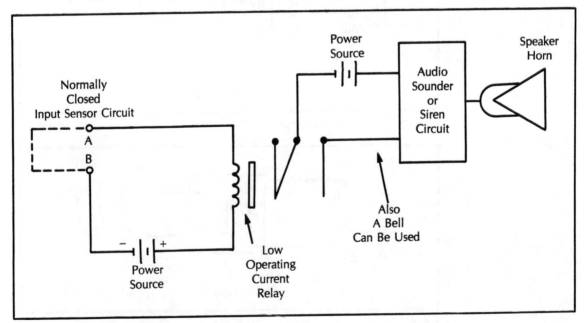

Fig. 4-4. Simple N.C. input alarm control system.

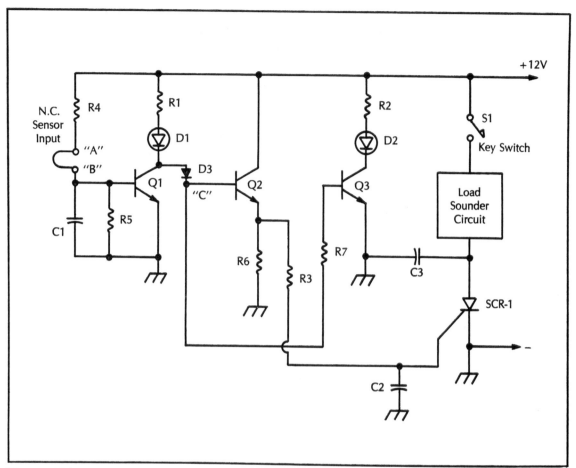

Fig. 4-5. N.C. alarm control system.

Table 4-1. Parts List for Fig. 4-5.

D3	IN4001 silicon diode
D1, D2	LED diodes, any color
C1-C3	.1 mF 100V mylar capacitor
Q1-Q3	2N3904 npn general purpose transistor
SCR-1	RCA S2800B, or similar 6 A device
R1, R2, R3	1 kΩ ½ W resistor
R4	2.2 kΩ ¼ W resistor
R5, R6, R7	10 kΩ ¼ W resistor
S1	Alarm set/reset key switch
Misc.	12V power source to supply current for whatever alarm sounder is used, cabinet, perfboard, pins, terminal strip, etc.

This can include all of the solid-state sensors that use relay or other mechanical contacts in the normally closed condition. All of the N.C. sensor contacts must be wired in series and then connected to the sensor input terminals to operate properly. A multi-terminal screw strip with the required number of slots can be connected to the A and B sensor input circuit of the control to make the job easier.

N.C. Alarm Control System Circuit Operation

The current flow through the sensor contacts and R4 supplies a forward bias to turn on transistor Q1. With all sensors normally closed, the LED D1 lights, indicating a good set for the alarm system. The voltage at the collector of Q1 is near ground level, keeping both Q2 and Q3 turned off. The LED D2 in the collector circuit of Q3 is also off. When a sensor opens, sending an alarm condition to the control unit, the forward bias to Q1 is removed and D1 goes out. The voltage at Q1's collector rises to near 12 volts and Q2 and Q3 are turned on.

The voltage at the emitter of Q2 supplies a current through R3 to the gate of the SCR, turning it on and supplying power to the alarm sounder circuit. Q3 lights up D2, indicating an alarm condition. Once the alarm has been activated, the power supplying the SCR must be interrupted to reset the system, and if a sensor remains in the alarm condition as the power is turned back on with the key switch, S1, the sounder goes again. If D1 is on and D2 off, the system can be reset without sounding the alarm, but if not, check out each of the sensor output terminals and make the appropriate correction before turning the system back on.

If a low-powered alarm sounder is used, the SCR can be replaced with a lower current device, or you can use the suggested one and have ample current reserve for expansion at a later date. If the sounder circuit operates with a complete off and on current demand, then to keep the SCR from turning off during the time the current demand is zero, a resistor should be connected across the SCR's load. The resistor value can be figured if the value of holding current is known for the SCR used in the circuit. For a holding current of 25 mA, the resistor value should be about 470 Ω in value. To figure the resistor value for any of the SCR's, just divide the holding current figure into the supply voltage and select the nearest standard resistor value.

A Power Supply

A suitable power supply can be built for the control unit with a battery backup to ensure operations when the power fails, or if the power is turned off by the intruder. The circuit in Fig. 4-6 can handle this and other control systems with similar power requirements. The current rating of the transformer can be determined by totaling the current required by each circuit powered by the supply. The largest current drain will be from the alarm sounder circuit.

The ac voltage at the secondary of T1 is rectified with four power silicon diodes in a full wave bridge circuit, and filtered with a 2000 μF electrolytic ca-

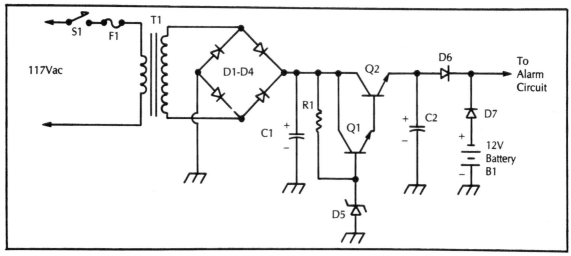

Fig. 4-6. Alarm power supply circuit.

pacitor. The raw dc voltage is fed to a two-transistor voltage regulator circuit using a zener diode as a voltage reference. Actually, for most control circuits the regulation is not needed, and the two transistors, zener, and R1 can be removed and a jumper placed where the collector and emitter of Q2 is located in the schematic.

By using the two diodes D6 and D7 the standby battery B1 automatically supplies power to the output when the ac power is cut. No power flows from the battery as long as the power supply is working.

Building the N.C. Alarm Control System

Using a metal cabinet for enclosing the circuitry adds an extra degree of class and security to the complete system without adding too much to the cost. Always start out with a cabinet somewhat larger than is needed to allow for

Table 4-2. Parts List for Fig. 4-6.

B1	12V battery used for back-up supply
C1, C2	2000 mF 25V electrolytic capacitor
D1, D2, D3, D4, D6, D7	3 to 6 A silicon diodes, size determined by circuit's requirement
D5	12 or 14V 1 W zener diode
F1	1 A fuse
Q1	2N2102 or similar npn transistor
Q2	2N3055 power npn transistor
R1	330 Ω ½ W resistor
T1	117Vac/12 or 14V, 3 to 6 A transformer
Misc.	Heat sink for both transistors, perfboard, pins, wire, cabinet, etc.

additional circuits that might be added at a later date. The same goes for building a power supply at least 25 to 50 percent larger than is needed for the basic system.

The complete control circuit can be built on perf board with the component parts mounted on push-in pins, or if you are set up for making printed circuit boards, make a layout and produce a p.c. board for the circuit. Unless you are going to build up more than one circuit, the perf board route is the fastest and as good as any to use. The power supply circuit can be built in the same manner, with the exception of the power transformer, and it should be firmly mounted to the bottom of the metal cabinet.

A heat sink can be made from a 2-inch square of scrap aluminum for the SCR, and the 2N3055 power transistor used in the power supply. The only time that the heat sink is needed is during the time the alarm sounder is activated, and under normal conditions the actual on time is short.

MULTI-INPUT ALARM CONTROL SYSTEM

The multi-input alarm control circuit is designed to match up with a number of different sensor circuits by using a separate input circuit for each sensor connected. The drawing in Fig. 4-7 is a multi-input interface circuit for sensors with N.C. outputs and is designed to connect up to the control system in Fig. 4-5, but can be used with other systems if connected properly.

Transistors Q1 through Q4 are connected as four independent switching circuits, and with a N.C. circuit connected to the input, operate like a switch keying the voltage at each of the collectors to ground, giving a zero output. The LED in each collector circuit lights when a N.C. circuit is connected to the circuit's input indicating a good set. If any of the input circuits changes to an OPEN circuit the forward bias is removed from the base of that transistor and its collector voltage rises from zero to near the supply voltage passing through one of the diodes D5-D8 to the point C of the control system in Fig. 4-5. The circuit in Fig. 4-5 then functions as originally designed by giving out an alarm signal. Each of the input circuits is similar to the one used in the control circuit of Fig. 4-5, but with diodes to direct and isolate each output.

Any number of input transistor stages can be added to the circuit in Fig. 4-7 to increase the number of sensor inputs needed for any installation. Just duplicate the parts used in a single stage and connect its output through a diode, like D5, to the output bus feeding point C of Fig. 4-5. For any of the input transistors not connected to a N.C. output sensor, a jumper can be connected in its place so an alarm output won't be sent out on an unused input.

If a number of normally open sensor circuits are to be used with your control unit, the interface circuit in Fig. 4-8 will fill the bill. Transistor Q1 is connected through two 4.7 k resistors and the normally open sensor switch to the positive power source. Resistor R2 and the bypass capacitor C1 offer a degree of ac isolation to the input stage. With a N.O. sensor connected to terminals A and B, Q1 is in a nonconducting state and Q2 is biased on with the current through R5. The LED D1 is on showing that a good set is made. No voltage is present at the output going to point C of the circuit in Fig. 4-5. If the sensor

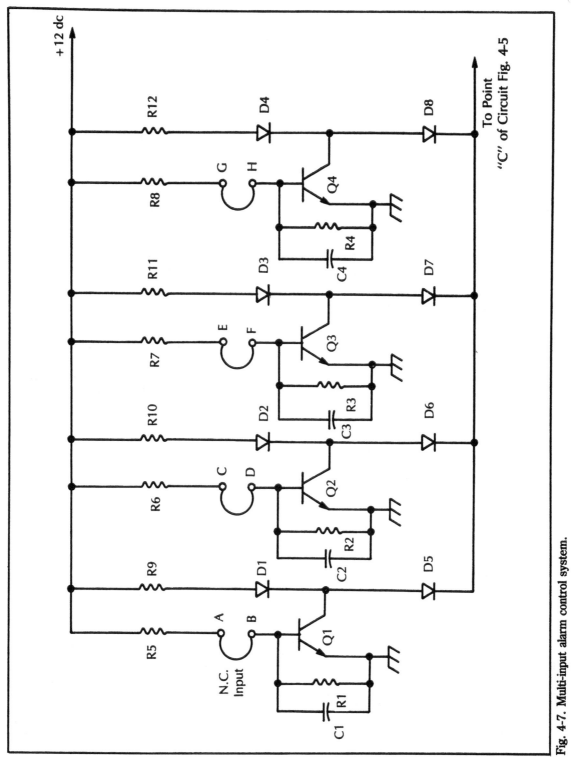

Fig. 4-7. Multi-input alarm control system.

Table 4-3. Parts List for Fig. 4-7.

C1-C4	.1 mF 100V mylar capacitor
D1-D4	LED, any color
D5-D8	1N4001 silicon diode
Q1-Q4	2N3904 npn general purpose transistor
R1-R4	10 kΩ ¼ W resistor
R5-R8	2.2 kΩ ¼ W resistor
R9-R12	1 kΩ ½ W resistor
Terminal strip for input circuits	
Misc.	Perfboard, pins, 12V power source, etc.

is triggered, a forward bias flows through the two resistors R1 and R2, and the sensor's closed contacts, to cause Q2 to be biased off and supply a positive voltage to C.

As the circuit is triggered, the LED goes out to indicate which of the sensors has been tripped, and the positive output sends an alarm signal to the control unit. The circuit shown in Fig. 4-8 is for two individual N.O. sensor inputs, but any number of sensors can be connected by duplicating one-half of the circuit as many times as is required.

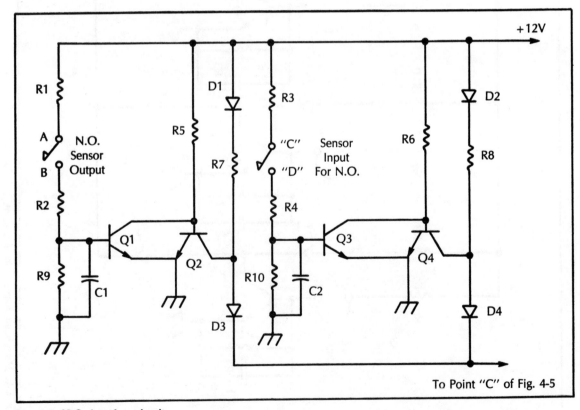

Fig. 4-8. N.O. interface circuit.

Table 4-4. Parts List for Fig. 4-8.

C1, C2	.05 mF 100V mylar capacitor
D1, D2	LED, any color
D3, D4	1N4001 silicon diode
Q1-Q4	2N3904 npn general purpose, or similar
R1, R2, R3, R4	4.7 k Ω ¼ W resistor
R5, R6	10 kΩ ¼ W resistor
R7, R8	1 kΩ ¼ W resistor
R9, R10	100 kΩ ¼ W resistor
Misc.	Perfboard, pins, terminal strip for sensor inputs, cabinet, etc.

Now, you can begin to see how quick the basic N.C. alarm control system of Fig. 4-5 can grow in size, and why it is a good idea to plan ahead by selecting a cabinet large enough for the completed system. Not only do the interface circuits require more room, but the selection of the sounder circuit and the size of the power supply, including a back-up battery, need to be considered. Room should be allocated for mounting the LED's on the front panel in a good visible location. Unless a completed system has been fully planned out, it would be best to assemble all of the circuits required, at least on paper, and then see how much room is needed to house the system.

IC ALARM CONTROL SYSTEM

A quad two-input nor gate IC is the heart of the N.C. alarm control system shown in Fig. 4-9. The CMOS IC was selected for the input circuitry because of its immunity to high level noise and its low power needs. This low-cost device can operate with supply voltages starting below 5 volts and going up to a maximum of 16Vdc. The alarm circuit is designed to operate best with a power source of 9 to 15 volts, but if a lower voltage operation is needed the component values can be altered to work down to a 5 volt level.

The circuit functions in the following manner. Each of the four input sensor circuits are wired identically with their inputs tied to ground through the normally closed sensor contacts. This produces a high voltage level at the output of each of the gate circuits, lighting the four LED's indicating a good set for each input. The output of each gate is connected to a common bus through a 1N914 silicon signal diode, and is normally in a high level mode supplying a forward bias to transistor Q1. With Q1 switched on its collector voltage is near zero, but when any of the sensors changes to an open circuit, the voltage rises at the collector, supplying a positive gate voltage to the SCR.

The SCR latches on, supplying current to the alarm sounder circuit, turning it on to sound off until the manual key switch S1 is reset or turned off. If the sensor that produced the alarm signal resets itself automatically, as a proximity circuit does, the key switch only needs to be turned off momentarily to break the current flow and then reset, ready for another sensor output.

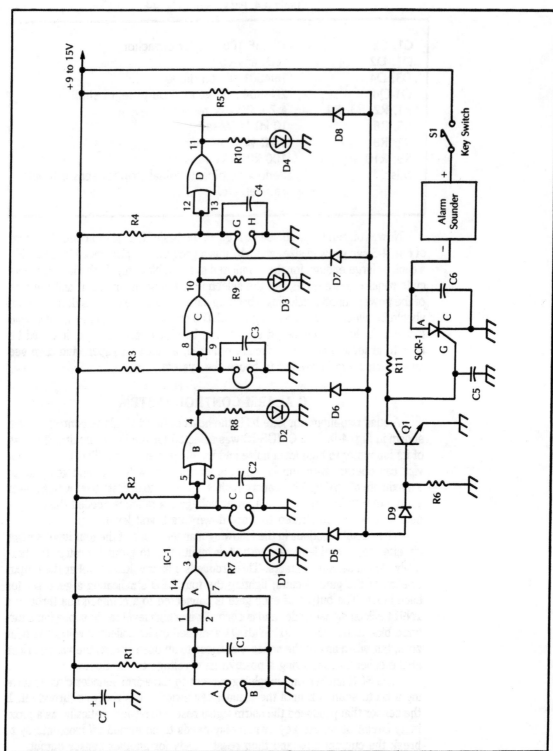

Fig. 4-9. IC N.C. alarm control system.

Table 4-5. Parts List for Fig. 4-9.

C1-C6	.1 mF 100V mylar capacitor
C7	220 mF 25V electrolytic capacitor
D1-D4	LED, any color
D5-D9	1N914 silicon diode
IC1	4001 MOS quad two-input nor gate IC
Q1	2N3904 npn or similar general purpose transistor
SCR-1	RCA S2800B, or similar 6 A device
R1-R6	15 kΩ ¼ W resistor
R7-R11	1 kΩ ½ W resistor
Misc.	Terminal strip for input sensor circuits, perfboard, pins, wire, cabinet, etc.

An ac operated power supply with a battery backup provides the circuit an uninterrupted source that keeps the alarm on guard full time, even during a short power outage. The basic power supply circuit in Fig. 4-6 would do for this circuit.

The IC control circuit can be assembled on perf board, or use any construction scheme that works best for you, as the component layout is noncritical, and as long as good acceptable wiring methods are used, no problems should occur. A socket should be used for the IC, and a 2×2-inch heat sink should be attached to the SCR. The IC contains internal circuitry to protect the inputs against damage from external static voltages, but it would be a good idea to not let any static or voltages higher than the supply be connected to the circuit in any way.

Placing the IC Control Unit in Service

A complete system must contain a control circuit, a power supply, an alarm sounder circuit, and a number of input sensors to send electronic messages of an alarm condition to the system. The sensors connecting to the IC circuit must be of the N.C. output type that open when an alarm signal is given. Any of the solid-state sensor output stages can be used if their outputs are designed to clamp a positive voltage to ground potential for a normal non-alarm output.

The interface circuit in Fig. 4-10 shows how easy it is to connect the sensor circuit in Fig. 3-24 to the IC control system. The output circuit of the vibration pick-up in Fig. 3-24 uses a NPN transistor as a switch to produce a N.C. output. To transfer this N.C. output in a completely isolated manner, an optocoupler with a NPN transistor output is used to repeat the N.C. output by switching the voltage at the input of the IC circuit to ground potential. An extra feature is obtained by using the optocoupler circuit; assuming the coupler is located at the control unit location, and if the two wires going to the control unit are cut or shorted together, an alarm signal is given out.

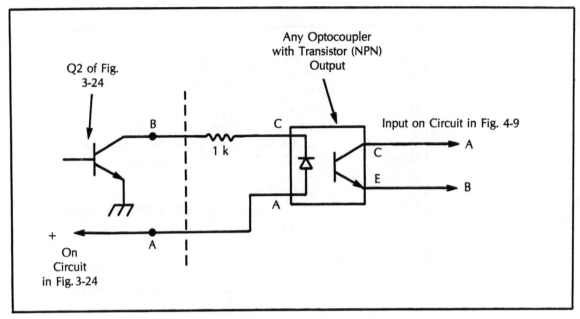

Fig. 4-10. Interface circuit.

In either condition, the current flowing through the LED diode in the optocoupler is stopped and the forward bias is removed from the internal transistor, allowing the output circuit to change to an open circuit sending out the alarm signal.

The IC control system can also operate with sensors that offer a normally open circuit output by using the interface circuit in Fig. 4-11. An optocoupler with a transistor output, similar to the one in the previous circuit, is direct coupled to a 2N3904 NPN transistor with a forward bias current flowing through a 10 k resistor. As long as the sensor is sending out a normal open circuit, the LED remains dark and the internal transistor is switched off, letting the bias current flow into the base of the 2N3904 transistor, turning it on and producing a N.C. output to the input of the control circuit. This circuit is the electronic equivalent of a relay that uses the N.C. contacts as the output, with the relay in a non-operated state. A sensitive relay can be used in this manner, as the reversing component for the sensor circuit, but the reliability offered by the solid-state circuitry is usually the best choice.

AN ENTRY/EXIT SENSOR DELAY CIRCUIT

The main entrance to a building or home can have its own time delay circuit to allow entry and exit without setting off the alarm. The advantage of a timed exit/entry circuit for a specific door is to allow the control system and other sensors to function normally without the delay affecting their operation. In most systems that have a delay feature, the alarm will not be triggered by another sensor as long as the system is on a timed hold. A smart burglar could find this out and make use of that fact by breaking in and destroying the control system

100

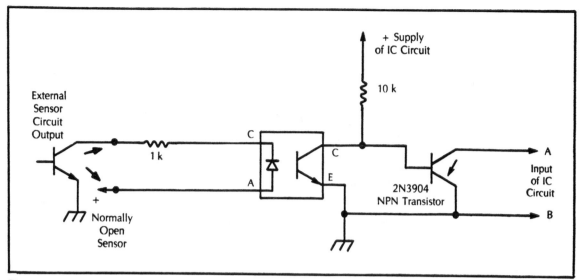

Fig. 4-11. N.O. interface circuit.

before the time delay has completed its cycle and sounds the alarm. If only the entry sensor is timed, the control system and remaining sensors operate in a normal manner to announce the break-in.

An entry/exit sensor delay circuit is shown in Fig. 4-12. Two reset switches are used, one for exiting and the other for making an entry without setting off the alarm. The switches shown are single pole push button switches, but for added security a number of paralleled switches can be used and coded to limit their use to authorized personnel only.

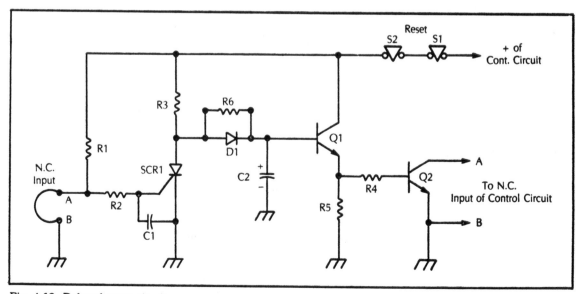

Fig. 4-12. Delayed entry circuit.

Table 4-6. Parts List for Fig. 4-12.

D1	1N4001 1 A silicon diode
C1	.1 mF 100V mylar capacitor
C2	47 mF 25V electrolytic capacitor
Q1, Q2	2N3904 npn transistor
SCR-1	Small .8 A 100V or greater low-current SCR
R1	2.2 kΩ ¼ W resistor
R2, R3	1 kΩ ½ W resistor
R4, R5	10 kΩ ¼ W resistor
R6	100 kΩ ¼ W gives 15-second delay,
	220 kΩ ¼ W gives a 45-second delay
S1, S2	N.C. push button switches
Misc.	Perfboard, pins, cabinet, wire, etc.

Delay Entry/Exit Circuit Operation

The door sensor is connected to the input terminals A and B of the delay circuit, keeping the gate voltage of the SCR-1 at ground potential. The timing capacitor C2 is charged up through R3 and D1 instantly when the power is turned on, to a voltage near that of the supply. This capacitor voltage is isolated from output loading by the emitter follower transistor Q1. The voltage at the emitter of Q1 is almost the same as the power source voltage, and it supplies forward bias for transistor Q2 to operate like a closed switch. This N.C. output connects to the sensor input of the control system.

If the door is opened by a burglar without knowledge of the delay system's reset procedure, the sensor opens and the current through R1 and R2 triggers the SCR on, starting the discharge of the timing capacitor through the timing resistor R6, to ground through the SCR. When the voltage across the capacitor drops to about one volt or less, transistor Q2 loses its forward bias and turns off, opening the sensor output circuit to give out an alarm signal. The only way to intercept the delayed alarm signal is to locate and activate one of the reset switches.

The actual time delay can be set by selecting the resistance value of R6. A 100 k resistor offers a delay of about 15 seconds, and a 220 k gives about 45 seconds of delay, but by experimenting with larger values the time can be extended to more than a minute.

The circuit can be constructed on perf board with push-in pins and located in the control unit's cabinet, or at the location of the input sensor. One of the reset switches needs to be located inside for entry reset, and the other outside the entrance to reset for exiting. The two reset switches should be mounted in an area not obvious to everyone, but placed where they are convenient to reach and activate.

The switch arrangement shown in Fig. 4-13 can be used to replace either or both of the reset switches to make it more difficult for someone to find and reset the time delay circuit. All five of the switches can be arranged in any order

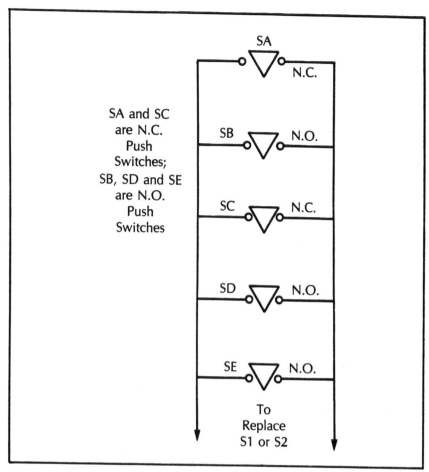

SA and SC
are N.C.
Push
Switches;
SB, SD and SE
are N.O.
Push
Switches

SA N.C.

SB N.O.

SC N.C.

SD N.O.

SE N.O.

To
Replace
S1 or S2

Fig. 4-13. Reset switch.

for convenience, and additional switches can be used to increase the difficulty in resetting the alarm without knowing the code.

Switches SA and SC are normally closed push button switches, and SB, SD, and SE are normally open push button switches connected as shown in Fig. 4-13 to set a code for resetting the timer circuit. Switches SA and SC must both be pressed at the same time to cause a current interruption to the SCR circuit, to let it turn off and reset the timer. If any of the three switches (SB, SD, or SE) happen to be pressed when SA and SC are activated, the circuit remains unbroken and the timer times out, setting off the alarm. The correct combination of SA and SC must be pressed at the same time to reset the circuit.

It doesn't matter too much which of the control systems you select to build and use, but it is very important how the complete system is installed. A poorly-placed sensor with exposed wiring, or any part of the system that's improperly installed, can give a burglar a clue how to circumvent the system and take the goods. The simplest system properly installed is many times more effective than the most sophisticated system installed in a sloppy or helter-skelter way.

Chapter 5
Alarm Indicators
and Audio
Sounders

Now that a selection of sensor circuits and control systems are at your disposal, a device is needed to convey the alarm message to the outside world. Basically there are three ways to deliver the alarm signal to its destination. The method most often used is to create a loud noise in the form of a bell or electronic siren that warns the burglar and at the same time tells the surrounding area that a burglary is in progress so the authorities can be called in. Another way to announce the crime is to turn on a number of lights as a warning to the burglar and to signal for the authorities. A third approach is the silent method, where the burglar is allowed to go on about his business while the alarm signal is sent by wire or radio to a central control station.

INDOOR ALARM INDICATORS

In many alarm installations a means of warning personnel inside that a breach of security is in progress is a necessary part of the alarm system. A perfect example is the homeowner who's in bed counting sheep when an intruder sets off an alarm sensor and sounds the outside alarm, but the sheep and sleep keep the master from hearing it. A low level audible sounder, with as many speakers as necessary to cover the desired area, is a good method to use to announce a break-in to the insiders.

The electronic sounder circuit in Fig. 5-1 is ideal for inside use. The output is a single continuous audio tone that can be set, with a frequency pot, to less than 100 Hz or to over 4 kHz, and is of a square-waveform in shape. A horn

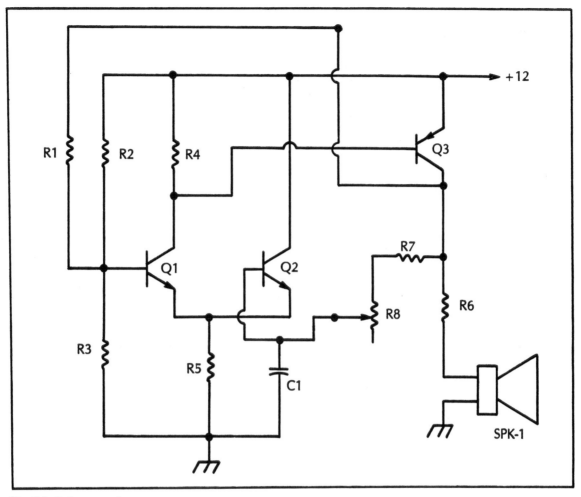

Fig. 5-1. Indoor sounder.

Table 5-1. Parts List for Fig. 5-1.

C1	.27 mF 100V mylar capacitor
Q1, Q2	2N3904 npn general purpose transistor
Q3	2N3906 pnp general purpose transistor
R1, R2, R3	15 kΩ ¼ W resistor
R4	10 kΩ ¼ W resistor
R5	3.3 kΩ ¼ W resistor
R6	100 to 200 Ω ½ W resistor
R7	1 kΩ ¼ W resistor
R8	100 kΩ pot
SPK-1	16 to 45 Ω speaker; horn type is best
Misc.	Perf board, pins, cabinet, wire, knob, etc.

speaker of 45 Ω matches the circuit best, but speakers with 16 Ω impedance work also. If two speakers are needed, connect them in series and hook to the output of the sounder circuit. If the sounder is to be used in a high-noise environment, the output of the circuit might not be high enough to overcome the surrounding noise, and an outside alarm sounder circuit should be used.

A SINGLE TONE GENERATOR

The operation of the single tone generator is very simple, with the three transistors connected in an astable multivibrator circuit. The tone frequency is determined by the values of C1 and the setting of R8. The output is taken off at the collector of Q3 with a current limiting resistor R6 that keeps the transistor from drawing too much current and being destroyed. If a 45 Ω speaker is used, R6 can be as low as 100 Ω, and if a 16 Ω speaker is used try a value of 150 to 220 Ω. If at any time Q3 shows any signs of heating, the value of R6 is too small, and should be increased.

The circuit uses very few parts that, if only one speaker is to be used, it can be constructed on a small section of perf board and possibly be mounted inside of the speaker's housing. If not, select either a small plastic or metal cabinet for housing the circuit in. In any case the construction can use any scheme that works out best, as the circuit is non-critical in its component layout. If more than one circuit is required, try your hand at making a p.c. board layout and make as many tone generators as you need for the installation.

AN INTERRUPTED TONE GENERATOR

The circuit in Fig. 5-2 offers an interrupted tone generator that produces a sound that's very hard to miss even in a high noise area. If you have ever been in an area where a piece of equipment produced a continuous hum, it usually only takes a short time to completely put it out of mind and not even be aware that a noise is present. Ever try putting an interrupted noise out of mind? It's almost impossible to do, and that's why this sounder works so well.

Two oscillators are combined to give a tone that is keyed on and off at about a 1 Hz rate. One half of a quad two-input nor gate operates as a low frequency generator circuit, producing a square-wave voltage at a 1 Hz rate at its output, pin 4. This switched on and off voltage is tied to one input of a nor gate in the second oscillator circuit that is operating at a frequency of about 1 kHz. This keys the 1 kHz oscillator on and off at a 1 Hz rate and the interrupted output appears at pin 11. A PNP transistor, Q1, drives the 45 Ω horn speaker with ample volume for a medium size shop or office. If more than one speaker is needed, two 16 Ω speakers can be connected in series, or, since the circuit is so simple and cheap to build, a circuit could be used at each speaker location.

The low frequency oscillator can be set to operate at either a higher or lower cycle rate by changing the value of R1 and C2. Increase either or both in value and the frequency goes down, or decrease either or both and the frequency rises. The same holds true for the higher frequency oscillator by changing the values of R2 and C3.

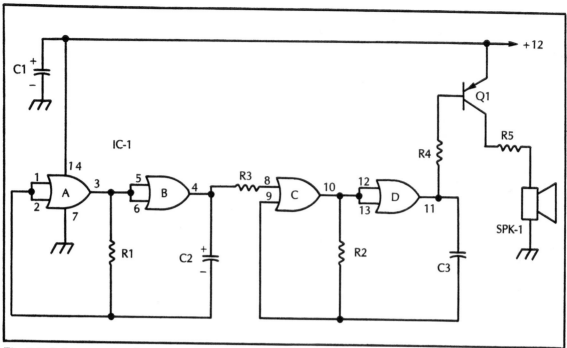

Fig. 5-2. Indoor sounder circuit.

Transistor Q1 is not a power transistor so don't lower the value of R5 below 100 Ω or the device could be destroyed. If the sounder is to be used in a small area where the noise level is low, a standard paper cone speaker can be used. To lower the output level experiment by using higher values for R5.

Any good construction scheme can be used in building the sounder circuit. The old standby procedure of perf board and push-in pins will do just fine, and the completed circuit can be housed in any cabinet of suitable size. If a horn speaker is used and there is enough room inside the driver enclosure, mount

Table 5-2. Parts List for Fig. 5-2.

C1	100 mF 16V electrolytic capacitor
C2	4.7 mF 25V electrolytic capacitor
C3	.01 mF 100V mylar capacitor
IC-1	4001 quad two-input nor gate IC
R1, R2	100 kΩ ¼ W resistor
R3	15 kΩ ¼ W resistor
R4	2.2 kΩ ¼ W resistor
R5	100 to 220 Ω ½ W resistor
SPK-1	45 Ω speaker horn
Q1	2N3906 npn general purpose transistor
Misc.	Perfboard, pins, wire, cabinet, etc.

the circuit inside. Just take normal care while working with the MOS IC and do not overexpose it to excess heat or high voltages.

A POWER ALARM SOUNDER

The high level output sounder in Fig. 5-3 is a modification of the low power circuit of Fig. 5-2. The output transistor is a 2N3055 power device that can drive the larger outdoor metal horn speakers. Either a pulsed tone or a two tone output can be selected, and the off/on rate can be increased approximately two-fold. The output tone frequency can also be increased by a factor of two. These special features are made available by the addition of a few resistors, three switches, and a power transistor.

Operation of the power sounder goes like this: Gates A and B make up the low frequency on/off oscillator, with the rate set by the values of R1, R2, and C2. With S1 in the normal open position, the rate is about 1 Hz, and increases to 2 Hz with S1 in the closed position. The square-wave output at pin 4 is connected to the wiper of a single pole two-position switch, S3. With the switch in the A position the signal is coupled to the input gates of section D of the IC, and produces a high/low frequency two tone output. When S3 is placed in the B position the on/off signal is coupled to an input gate of section C, turning the output oscillator on and off at the same rate.

If only a fixed on/off pulse rate is desired a 100 k pot can be used to replace R1, and R1, R2, and S1 can be eliminated from the circuit. The same can be

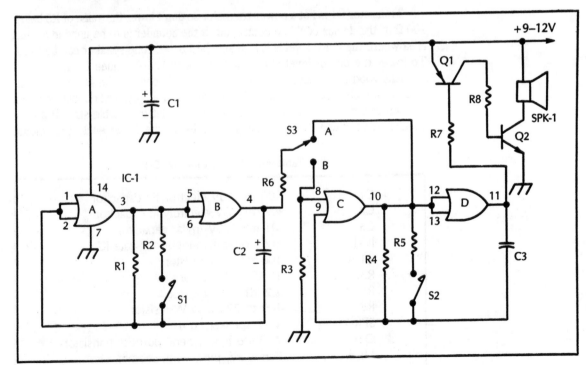

Fig. 5-3. High power sounder.

Table 5-3. Parts List for Fig. 5-3.

C1	100 mF 25V electrolytic capacitor
C2	4.7 mF 25V electrolytic capacitor
C3	.01 mF 100V mylar capacitor
IC-1	4001 quad two-input nor gate IC
Q1	2N3906 pnp transistor
Q2	2N3055 pnp power transistor
R1, R2, R3, R4, R5	100 kΩ ¼ W resistor
R6	3.3 kΩ ¼ W resistor
R7	2.2 kΩ ¼ W resistor
R8	100 Ω ½ W resistor
SPK	Speaker horn for outdoor use, 16 Ω
S1, S2	SPST toggle or slide switch
S3	Single pole two position switch
Misc.	Wire, cabinet, perfboard, pins, etc.

done for the tone output generator. A 100 k pot can replace R4, and R4, R5, and S1 can be removed. Both pots can be set for the desired effect while the circuit is operating, but you will probably want to aim the speaker outside, or connect a 470 Ω resistor in series with the speaker to lower its output.

If a number of different locations are to be covered with a single circuit, and they are separated by more than a few hundred feet, it is best to build individual sounder circuits for each location. Two locations within one hundred feet of each other can be handled by a single circuit with two 8 Ω speakers connected in series, but the power output is divided between the two locations. If more output power is required, then a circuit for each speaker is the logical answer.

The circuit can be built like the previous one on perf board and, if room is available, mounted in the speaker's case. The circuit should be located as close to the speaker as possible to reduce the power loss due to long wire runs.

AN AC LIGHT CONTROLLER

In some instances a burglar can be discouraged, if the entry is made during darkness, by turning on a group of lights to illuminate the break-in area. A nighttime burglar is like a fish out of water when he or she is caught working in a lighted area. The first instinct the burglar has when the light is turned on, is to flee back into the world of darkness. Any time a burglar cannot be caught, the next best thing to have happen is to scare him away with the memory of the near catastrophe etched in his mind forever.

The circuit in Fig. 5-4 takes the control system's output and converts it into a signal that can turn on a group of ac lamps. The current required to drive the light circuit is very small, so the audible alarm sounder can also be operated at the same time if it is desired. In either case the lights remain on until the control system has been manually reset with the key switch.

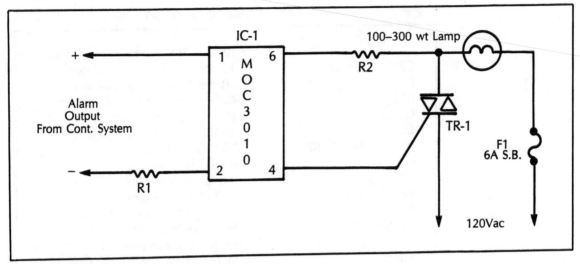

Fig. 5-4. Visual indicator.

A MOC3010 optocoupler triac driver IC receives the alarm sounder's drive voltage when the alarm control system is activated. This voltage supplies current to the internal LED that light-couples the information to the silicon bilateral switch that, in return, supplies gate current to the 6 A triac to turn on the lamps.

If a higher light output is desired, use a triac with a greater current rating, but be sure to check the gate current requirements to be sure that the drive circuit will supply ample gate current. Any triac with a gate current sensitivity of 50 mA or less works in the circuit. Also, select a fuse rating that offers protection to the triac used, and always have the 120Vac turned off and unplugged before working on the circuit.

The light control circuit can be constructed on a section of perf board and housed in a small cabinet. Since the component cost is low, a separate circuit can be built for each group of lights used, and each of the input circuits paralleled and connected to the control system's output.

AN INTERRUPTED AC LIGHT CONTROLLER

Since an interrupted audio tone is more effective in attracting attention than a single continuous tone, it would seem that the same effect can be expected

Table 5-4. Parts List for Fig. 5-4.

IC-1	MOC3010 optocoupler triac driver IC
TR-1	6 A 400V triac
R1	470 Ω ½ W resistor
R2	180 Ω ½ W resistor
Misc.	Perfboard, power cord, fuse holder, fuse, 100 to 300 W lamp, etc.

for a similar visual light display. If you have ever been exposed to a low frequency high wattage flashing light you know that, without closing your eyes, there is no way to ignore the effect. A light that's steady can be used to see our way around, but ever try to do anything when the only light available to see by is flashing on and off?

A light flashing circuit is shown in Fig. 5-5 that grabs the attention of anyone who is in visual range. The actual lamp control circuit is a duplication of the one in the previous circuit with an optocoupler controlling the triac. A timer circuit is added between the control system and the optocoupler to turn on and off the lamp at approximately a 1 Hz rate.

Here's how the circuit functions. When the control system sends out the alarm signal, approximately 12Vdc is connected to the input of the flasher circuit. A 555 timer IC is connected in a low frequency square-wave generator circuit operating at a frequency of about 1 Hz. The on and off time, at the output of the IC pin 3, is about equal in time. This supplies a pulsing current to the LED in the optocoupler that turns on the internal bilateral switch that supplies gate current to the 6 A triac.

The on/off timing rate can be varied by changing the value of C1 or R1, or both. If R1 and C1 are raised in value the frequency is lower. To increase the frequency just lower the value of either or both components.

AN ELECTRONIC SIREN

Another excellent attention-getter that everyone is familiar with is the two-tone electronic siren. To some the cry means the life-saving sound of an am-

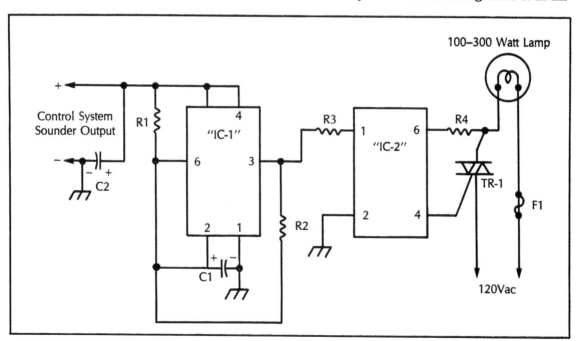

Fig. 5-5. Pulsed visual indicator.

Table 5-5. Parts List for Fig. 5-5.

IC-1	555 timer IC
IC-2	MOC3010 triac driver IC
R1	47 kΩ ¼ W resistor
R2	4.7 kΩ ¼ W resistor
R3	680 Ω ¼ W resistor
R4	180 Ω ¼ W resistor
F1	6 A fuse
TR-1	6 A 400V triac
C1, C2	47 mF 25V electrolytic capacitor
Misc.	100 to 300 W 120Vac lamps, power cord, perfboard, fuse holder, cabinet, etc.

bulance coming to the rescue, and to others, that the jig is up, get out or get caught.

The siren circuit in Fig. 5-6 gives any burglar the audible signal to clear out or face the music. The heart of the circuit is two 555 timer ICs that are connected in two separate oscillator circuits. IC-1 is generating a very low frequency tone of approximately 1 Hz, and IC-2 produces a shifting tone centered around 1.5 kHz. Pin 5 of IC-2 is modulated with the square-wave signal coming from the output of the 1 Hz oscillator, pin 3 of IC-1. A shifting frequency siren-like output is the result of the 1 Hz modulation.

The varying frequency output of IC-2 is coupled to the base of Q1 through two level shifting diodes and a current limiting resistor. The collector of Q1 supplies a signal current to the power transistor Q2 that drives the outdoor speaker.

The siren circuit is an easy one to modify to produce a number of different sounds. The rate of the up and down siren sound can be increased or decreased in rate by changing the value of R1. To lower the rate, increase the value, and to speed up the rate, decrease the value. The depth, or modulation level, is set by the value of R3. To increase the modulation level, lower the value, and to decrease the level, increase the value.

The center frequency of the second oscillator is set by the value of R4, and can be changed in the same manner as the low frequency oscillator. The timing capacitor used in each of the oscillator circuits can also be changed to vary the output frequency. As the capacitance is made larger the operating frequency goes down, and as the value of capacitance decreases the frequency increases. The timing capacitor for IC-1 is C2, and for IC-2 is C3. If the timing resistors are changed in value, then the value of the feedback resistor must also be changed along with it. The feedback resistor, R2 for oscillator 1, and R5 for the second oscillator, needs to be kept at a 10:1 ratio of the timing resistor. This ratio helps keep the square-waveform more uniform in its on and off timing cycle.

The construction of the siren circuit can follow any suitable scheme, as the component layout is non-critical, but unless a p.c. board is used, the IC's should be mounted in sockets. The power transistor can be mounted on a 3×3-inch

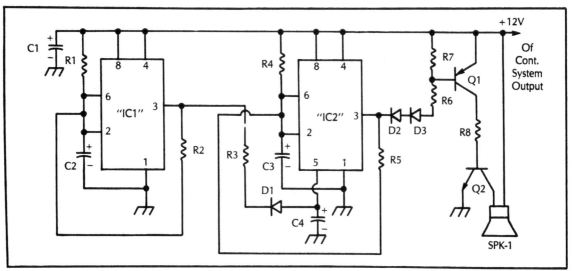

Fig. 5-6. Two-tone siren.

Table 5-6. Parts List for Fig. 5-6.

D1-D3	1N914 signal diode
IC-1, IC-2	555 timer IC
Q1	2N3906 pnp transistor
Q2	2N3055 power npn silicon transistor
C1	100 mF 25V electrolytic capacitor
C2	47 mF 16V electrolytic capacitor
C3	.5 mF 25V electrolytic capacitor
C4	4.7 mF 25V electrolytic capacitor
R1	47 kΩ ¼ W resistor
R2	4.7 kΩ ¼ W resistor
R3, R4	10 kΩ ¼ W resistor
R5, R6	1 kΩ ¼ W resistor
R7	3.3 kΩ ¼ W resistor
R8	100 Ω ½ W resistor
SPK-1	16 Ω horn-type speaker
Misc.	Perfboard, pins, wire, heat sink material, IC sockets, etc.

piece of aluminum for heat sinking, and the complete circuit mounted in a metal or plastic cabinet.

All of the visual and audible alarm circuits are designed to work with any of the control systems of the previous chapter. A sounder circuit and a lamp driver can both be connected in parallel to operate together from the single output of the control system. This arrangement can give your alarm system a double barrel shot to the first burglar that makes the mistake of hitting on your turf.

113

Chapter 6
Telephone Remote Monitoring Control System

The telephone that you use daily to order groceries, check on the price of a stock, or for the pleasure of hearing a loved one's voice, can also connect you to another phone circuit in almost any country on the face of the earth. But even more important is the fact that you can call your number from any of the millions of telephones anywhere in the world and instruct your phone-controlled circuit to relay information to you thousands of miles away, without needing anyone at your home location to assist in any way.

With the proper electronic circuitry connected to a standard telephone pair (TIP and RING terminals of the phone lines), an incoming call can be answered before the first ring has completed its cycle. All that's necessary to do to answer the ring is to bridge the telephone lines with a resistive load of approximately 600 Ω. As long as the load is connected to the pair, the phone circuit remains complete and information can pass to and from the bridged circuit.

A RINGER LATCH CIRCUIT

The block diagram in Fig. 6-1 can help to explain the operation of a telephone remote monitoring system. A ringer latch circuit is connected to the TIP and RING terminal of the incoming phone line, as well as the bridge and audio matching circuitry. When the ringing voltage is applied to the phone lines, the latch circuit brings the power up for all of the circuits and starts timer A, which keys on the phone line bridge circuit. This produces an answer condition that tells the phone company that a good bridge has been completed, and the line

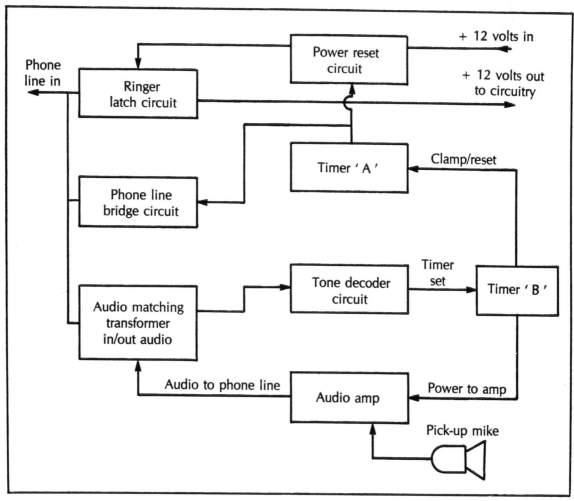

Fig. 6-1. Block diagram of remote telephone monitoring system.

is ready for sending and receiving audible information. After a set time period, timer circuit A cycles out unless a reset or clamp signal is received.

If timer A is allowed to time out without receiving the clamp signal, the line sounds dead to the calling party, and within a few seconds automatically disconnects and returns the dial tone to the caller. If the calling party sends the correct audio tone down the phone lines, the tone decoder circuit sends out a set signal to timer B. This timer performs a dual function by bringing the power up to an audio amplifier, and sending a clamp or reset signal to hold timer A activated.

The audio amplifier is connected to a mike that picks up sounds in the room or area and feeds that audio through a matching transformer down the phone line to be monitored by the calling party. When timer B times out, the amplifier goes dead and if the audio tone is repeated within about 5 or 10 seconds, the timer is reset, activating the amplifier again for the preset time period. If the

115

tone is not received within the allotted time period, the two timers cycle and the circuit is disconnected from the phone circuit. If the audio tone is repeated every 20 to 30 seconds, the amplifier can be kept on indefinitely.

A ZERO CURRENT STANDBY CIRCUIT

A special feature designed into this monitoring system allows the circuit to draw zero current from its power source during the time it is in standby waiting for a call. This feature allows the circuit to be self-contained with a battery power source, making a short term installation a snap. If a permanent installation is desired, an ac-operated power supply can be used.

Take a look at the circuit in Fig. 6-2 to see how the zero current standby feature is accomplished. A neon lamp is connected in series with a full wave diode bridge rectifier circuit. As long as the voltage across the phone lines remains at a normal 48Vdc level, no current flows through the circuit because the neon lamp takes approximately 60 to 70V to fire and turn on. When the phone rings, the ac ringing voltage level is sufficient to light the neon lamp to connect the ac source to the bridge rectifier circuit. The dc output of the rectifier is connected to the input of a IC optocoupler that activates an internal triac that latches on. This switches the power on through the triac, setting the circuit in operation.

At the same time the power is keyed up with the triac in the optocoupler, timer A is started to latch a bridging resistor across the phone line to keep the phone circuit open for use. The optocoupler's triac remains on until either the battery goes dead or the current path is broken. Turning off a triac can some-

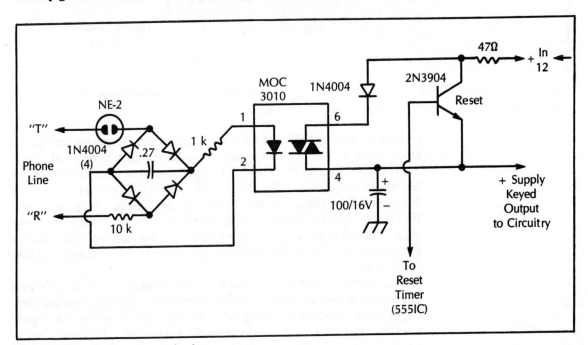

Fig. 6-2. Zero standby current circuit.

times be more difficult than getting one to activate. Basically there are two methods that can be used to turn off a scr or triac without using a hammer. The first and simplest method is to open the current path through the device for a short period of time. The second method is to divert the current around the device for a short period of time. Both methods cause the current flow through the triac to stop, turning it off until it is reactivated.

A reset transistor is connected across the triac, through a silicon diode, and when the A timer times out, a positive voltage is routed to the base of the transistor, turning it on and diverting the triac's current through it, turning off the triac. The silicon diode in series with the transistor and the triac offers a voltage offset that makes the job of turning off the triac with a silicon transistor more reliable. Now that we have gone over the basic function of the remote monitoring system shown in the block diagram, let's take an in-depth look at each of the working circuits.

THE BIG PICTURE

The complete circuit of the telephone remote monitor is shown in Fig. 6-3. IC-2 is timer circuit A, and IC-4 is B. IC-1 and IC-3 are the two optocouplers, and IC-5 is the PLL tone decoder chip. IC-6 and IC-7 are two op amp IC's that amplify the audio fed to the calling party.

THE PHONE MONITORING CIRCUIT

The A timer circuit is as good as any to start with, since the latching circuit has just been covered. The 555 timer IC is usually activated by a negative pulse fed into pin 2 of the IC, but in this timer circuit the trigger input is tied to ground through capacitor C13. When the power is turned on to the timer circuit, the trigger input is kept at ground level for a short time until C2 charges up to the supply voltage. This forced negative pulse starts the timer circuit, bringing its output (pin 3) to a positive voltage near that of the supply source and lighting the LED D11. Also, transistor Q6 is biased on operating the bridging relay, Rly-1.

The timing capacitor C1 and resistor R4 set the time period for IC-2. To increase the time delay, the two component values need to be made larger, and to decrease the time, the opposite. Transistor Q2 is connected across the timing capacitor to clamp the voltage to ground level to hold the timer in the activated condition until Q2 turns off. R6 limits the discharge current through the transistor to a safe level.

Optocoupler IC-3 drives the shut-down transistor, Q1. The anode of the internal LED is tied to the positive supply through a 1 k current limiting resistor, R8, and the cathode is connected to the output of timer A through a R/C combination of R9 and C3. As long as the timer is activated, the voltage at its output, pin 3, is about the same as the power source, and the voltage across the optocoupler's LED is near zero. When the timer times out, the voltage at pin 3 drops to zero, and the charging current through C3 also flows through the LED. The LED turns on the light-sensitive transistor, supplying a forward bias to the reset transistor Q1. Switching Q1 on shunts the current around the triac in IC-1, turning off the power to the monitor circuit.

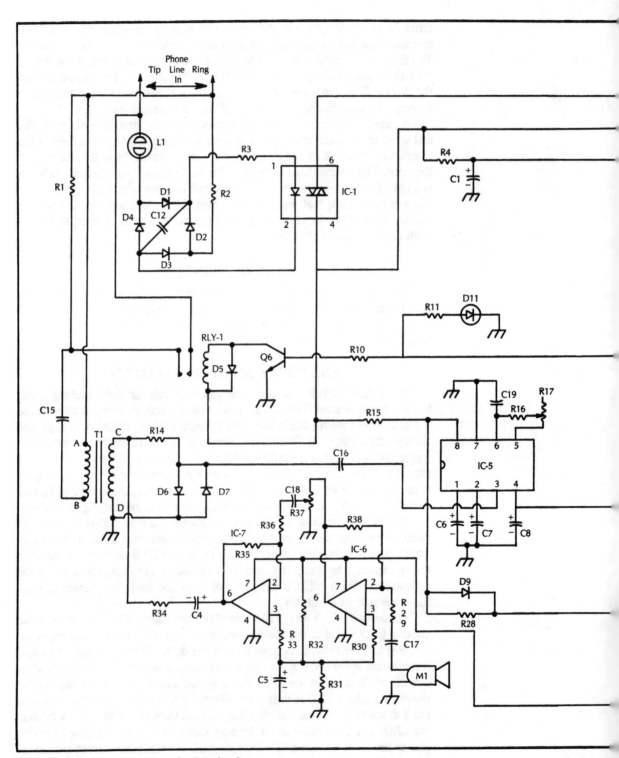

Fig. 6-3. Telephone remote monitoring circuit.

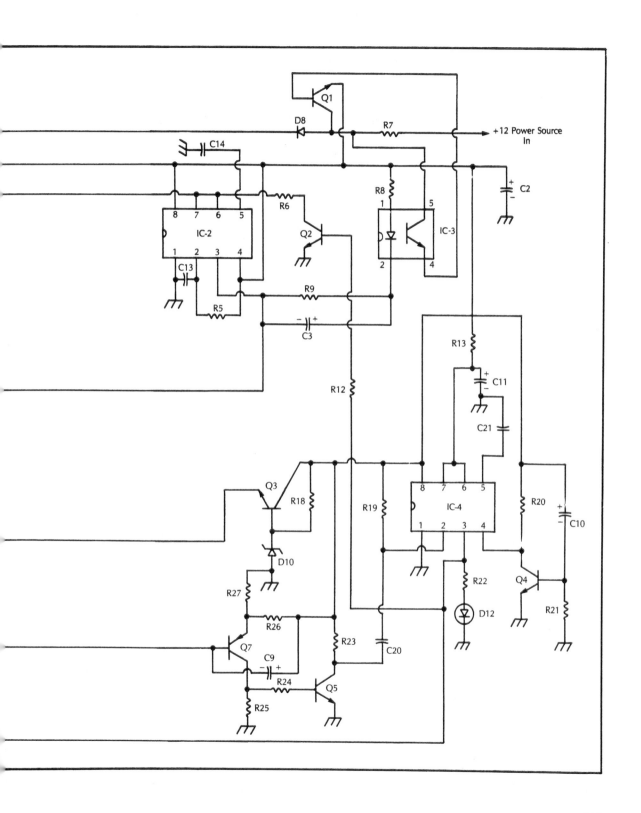

119

Table 6-1. Parts List for Fig. 6-3.

C1, C3, C4, C6, C9 C10	4.7 mF 16V electrolytic capacitor
C2, C5, C11	47 mF 16V electrolytic capacitor
C7	3.3 mF 16V electrolytic capacitor
C8	10 mF 16V electrolytic capacitor
C12, C15	.27 mF 100V mylar capacitor
C13, C17, C19	.12 mF 100V mylar capacitor
C14, C21	.01 mF 100V mylar capacitor
C16, C18, C20	.1 mF 100V mylar capacitor
D1-D5, D8	1N4004 lamp silicon diode
D6, D7, D9	1N914 silicon signal diode
D10	½ W 6V zener diode
D11, D12	LED, any color
IC-1	MOC 3010 IC, optocoupler
IC-2, IC-4	555 timer IC
IC-3	MOC 1000 optocoupler
IC-5	567 PLL IC
IC-6, IC-7	741 op amp IC
Q1-Q6	2N3904 or 2N2222 gp npn transistor
Q7	2N3906 or 2N3638 gp pnp transistor
M1	Small dynamic mike
Rly	Telephone-type sensitive 12Vdc, or other similar relay
R1, R3, R8, R10, R11, R15, R20, R22	1 kΩ ½ W resistor
R2, R12, R25, R29, R30, R33, R36	10 kΩ ¼ W resistor
R4, R35	470 kΩ ¼ W resistor
R5	22 kΩ ¼ W resistor
R6	100 Ω ¼ W resistor
R7	47 Ω 1 W resistor
R9	15 kΩ ¼ W resistor
R13	120 kΩ ¼ W resistor
R14, R16, R24, R31, R32	4.7 kΩ ¼ W resistor
R17, R37	20 kΩ pot
R18, R23	2.2 kΩ ¼ W resistor
R19	82 kΩ ¼ W resistor
R21, R26, R27	3.3 kΩ ¼ W resistor
R28	39 kΩ ¼ W resistor
R34	47 Ω ¼ W resistor
R38	680 kΩ ¼ W resistor
T1	600 Ω to 600 Ω line matching transformer
L1	NE-2 neon lamp
Misc.	IC sockets, perf board, pins, cabinet, battery holder, relay socket, wire, etc.

While timer A is activated, power is also turned on to the PLL tone decoder, IC-5, with transistor Q3 functioning as a 6V regulator. The audio coming in on the phone line is passed through T1 and a peak-to-peak voltage limiter circuit to the input, pin 3, of the PLL detector IC. The decoder's frequency can be set to any desired frequency within the response range offered by the phone system, with an adjustment of pot R17. When the correct tone is received by the PLL circuit the output of the IC-5, pin 8, is pulled to ground potential.

The base of Q7, a pnp transistor, is connected to the output of the PLL and is biased on when the decoder detects a tone. An npn transistor Q5, is direct-coupled through a current limiting resistor R24 to the output of the pnp transistor. As Q5 turns on, a negative pulse is coupled from its collector through C20 to the input of IC-4 timer B. The main purpose of the two-transistor circuit is to keep any noise or fast voltage pulses coming from the output of the PLL detector from falsely triggering the timer circuit. With the component values given in the parts list, the minimum tone time required for the PLL to respond is about 1 to 2 seconds.

An output from the tone decoder is the only way that timer B can be activated to turn on the audio amplifier and to clamp the timing circuit of timer A. LED D12 is on during the timing cycle. The timer-on time is set by the values of C11 and R13, and can be increased or decreased in the same method described for timer A. The time delay is about 20 seconds with the component values given in the parts list. Transistor Q4 is connected to pin 4 of the timer IC (reset input) to keep the timer from triggering on during the time that the power is keyed on. Capacitor C10 turns Q4 on during the time it is charging up to the supply voltage, and the collector of the transistor clamps the reset input of the timer to ground during the supply voltage transition.

The two-op amp audio amplifier is powered on when timer B is in its timing cycle. The input of IC-6, a 741 op amp, is connected to a sensitive dynamic mike, and its output feeds the volume level pot R37. The second op amp, IC-7, increases the audio level to drive T1 and the phone line.

If a battery supply is not used, be sure to select a good quality ac regulated 12Vdc supply with low ripple for the monitor circuit. The supply should be capable of a current output of 100 mA minimum.

Building the Telephone
Remote Monitoring Control Circuit

A good method to use in the construction of the monitoring circuit is to mount the components on a piece of perf board 6×8 inches in size and follow the general layout shown in Fig. 6-4. IC sockets should be used for the seven ICs, but the remaining parts can be mounted with push-in pins.

Start construction by locating the seven IC sockets in the general area shown in Fig. 6-4, and follow by placing the relay and matching transformer on the board. Push-in pins can be located and placed on the perf board for the seven transistors. This makes wiring the complete circuit easy without soldering the transistors in place until the job is near completion. The ICs can also remain out of the sockets until the wiring is completed. The actual component layout is not

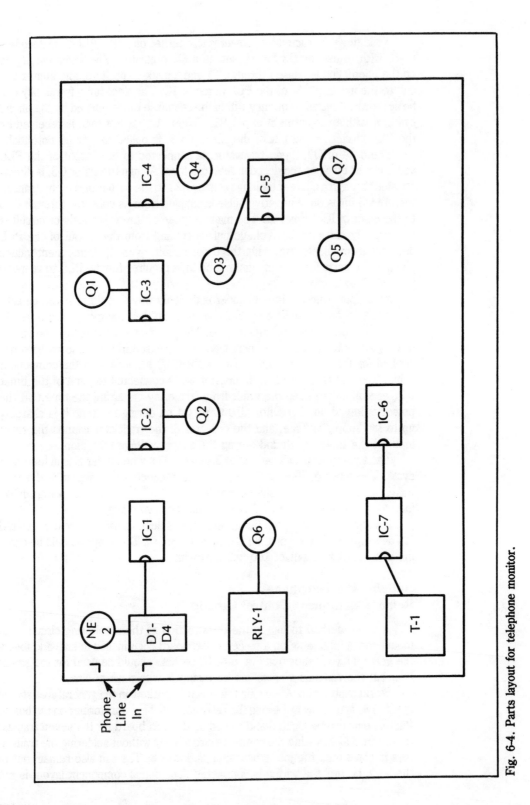

Fig. 6-4. Parts layout for telephone monitor.

critical, but by following the general layout the job of interconnecting the circuits is much easier.

A metal or plastic cabinet of suitable size can be used to house the circuit in. The two LEDs can be mounted on the front, along with the amplifier's gain control and the mike's input jack. If the unit and pick-up mike are to be located in the same area, the mike can be built in the cabinet. A long length of telephone wire with a molded plug can be wired to the circuit to allow you to locate the unit in the best place for picking up the desired sounds.

Placing the Monitor in Service

Set both pots to mid-position and connect a 12 Vdc source to the circuit. If everything is okay, nothing will happen and zero current will flow into the circuit. For the following tests and circuit adjustments, do not connect the unit to the phone circuit.

Connect one end of a jumper lead to pin 2 of IC-1 and the other to circuit ground. Connect another jumper lead to the plus side of the 12V power source, and temporarily touch the other end of the jumper to the positive output of the bridge rectifier (to the cathodes of D1 and D2). The bridging relay should activate and the LED D11 should light. After the timer circuit cycles out, the relay and LED turn off.

If the circuit performs as stated, the A timer is operating, optocouplers IC-1 and IC-3 are doing their thing, and transistor Q1 is resetting the triac in IC-1 to turn off the power. Disconnect the jumper leads.

Connect a jumper between pins 4 and 6 of IC-1 to bring up the power for all of the circuitry. Timer A and the bridging relay cycle for the A time period and return to the off condition. Take another jumper and connect one end to ground and touch the other end to pin 8 of IC-5. The second LED D12 should light indicating that timer B is activated and working. Connect a pair of high impedance earphones to terminals A and B of transformer T-1, and a mike to the input circuit, and again touch the grounded jumper lead to pin 8 of IC-5, and the amplifier should function.

If these few simple test procedures went as expected, then the overall circuitry is functioning as designed, and the next step is to determine what is needed for the tone transmitter and to set the PLL to its frequency.

TONE TRANSMITTERS

Two types of tone transmitters can be used with the circuit to activate the PLL decoder and turn on the monitoring amplifier. A solid-state battery-operated tone generator could be built with an internal mini-speaker to couple the sound to the phone's mouth-piece, but there is an easier electronic solution to the problem. A piezo-electronic sounder, a battery, and a push-button switch in a small plastic case make a super transmitter. The Mallory Sonalert solid state signal device has been around for years, and a number of these are available in miniature size with a choice of output frequencies. Radio Shack also sells a number of the solid state piezo sounders.

The simplest of all transmitters is the dime store whistle. A good quality whistle works just as well as either of the two electronic solutions without the bother of batteries or a device that fills a shirt pocket. The best quality single tone frequency whistle made of metal is a good choice, and the cost should be no more than the cost of a battery to operate the electronic tone circuit.

Connect the monitor to the phone lines, being sure that you are tied across the TIP and RING terminals. To make sure, take a dc voltmeter and check to see that approximately 50V is present across the lines when the phone is on hook. Hook the jumper back across pins 4 and 6 of IC-1 to bring the power up and lift the receiver off hook. You may need to dial part of a phone number to make the dial tone disappear so you can transmit the tone through the mouthpiece mike into the monitor circuit. With the tone feeding the phone, slowly tune the PLL with R17 until timer B turns on, lighting the LED. Pick up the handset, listen to the audio coming out of the two-stage amplifier, and set the gain pot R37 for the desired output.

Remove the jumper, go to a neighbor's house and call home. As soon as the ringing stops, send the tone. The amplifier should come on letting you hear what's going on at home. The usefulness of this electronic monitoring device is limited only by your imagination.

Chapter 7
Specialized Electronic Detection and Alarm Circuits

The majority of the electronic alarms in use in our high tech world today do not have anything to do with burglars or crime, but are primarily used to monitor high-dollar equipment and to give warning before a failure or breakdown occurs. Electronic and mechanical devices are the main types of equipment that depend on the safeguards that the electronic detection and alarm circuits can offer.

A total electronic system can fail due to an over or under supply voltage condition, and even if no damage occurs, the loss of information, as in a computer system, can be a costly experience.

OVER-VOLTAGE ALARM

A very simple but useful over-voltage alarm circuit is shown in Fig. 7-1. It can be adapted to a large number of protection applications. Almost any electronic circuit that's operated with a dc power source can use the circuit in Fig. 7-1, or a modified version of it, to monitor and send out an alarm when the voltage goes dangerously high. The circuit can be set to detect a voltage increase of a few millivolts or several volts, depending on the circuit's adjustment and the requirements of the protected equipment.

Here's how the circuit operates. A reference voltage is set by a zener diode, D1, that maintains a fixed voltage at the emitter of transistor Q1. The base of Q1 is connected to the wiper of a 10 k pot that is tied across the power source. As long as the voltage at the base is the same or lower than the voltage at the

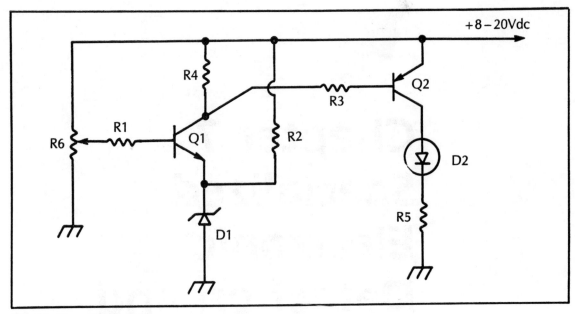

Fig. 7-1. Over-voltage alarm.

emitter, the transistor remains in the off state. When the voltage rises at the transistor's base to approximately .6V above that of its emitter, the transistor turns on. This supplies a forward bias to transistor Q2, turning it on to light the LED.

The basic circuit can be modified to do additional service by adding a relay, or SCR and relay, to not only give a warning of an over-voltage condition, but to shut down the equipment until the problem can be corrected. A SCR and relay can be easily added to the circuit by connecting the gate of the SCR to the top of R5, through a current limiting resistor for the device used, and the cathode to ground, with the relay connected between the SCR's anode and the

Table 7-1. Parts List for Fig. 7-1.

D1	5 or 6V zener diode
D2	LED, any color
Q1	2N3904 or 2N2222 npn transistor
Q2	2N3906 or 2N3638 pnp transistor
R1, R2, R3	2.2 kΩ ¼ W resistor
R4	10 kΩ ¼ W resistor
R5	1 kΩ ¼ W resistor
R6	10 kΩ pot
Misc.	Perfboard, cabinet, etc.

positive power source. The SCR latches the relay on, and maintains the condition until the circuit is turned off or the current path through the SCR is broken.

Setting up the circuit is no problem. If the power source that's to be monitored is a variable type, just connect the circuit across the supply and set the power supply's voltage to the over-voltage setting that you want the alarm to indicate. Then adjust pot R6 until the LED lights. Connect an accurate voltmeter across the power supply and lower the voltage until the LED goes out. Slowly increase the voltage of the power supply until the LED just lights, noting the meter reading at the same time. For power sources that only produce a fixed output voltage, adjust R6 until the LED just lights and then back off until the LED just goes out.

OVER/UNDER VOLTAGE ALARM

An over/under-voltage alarm circuit is shown in Fig. 7-2. A single op amp, of a quad op amp, is used in a voltage comparator circuit to indicate either an over or under voltage condition. One advantage this alarm has to offer is that a single circuit gives the choice of monitoring an over-voltage or under-voltage condition by how the pot R6 is adjusted.

The circuit operation is simple. The op amp has each of its inputs connected to a voltage source. The negative input, pin 2, is connected to a fixed reference voltage by zener diode D1. The positive input of the op amp, pin 3, is connected to the voltage adjustment pot R6.

When R6 is adjusted to where D3 just lights, the circuit is set to indicate an over-voltage condition. If the supply voltage creeps up ever so slightly, the

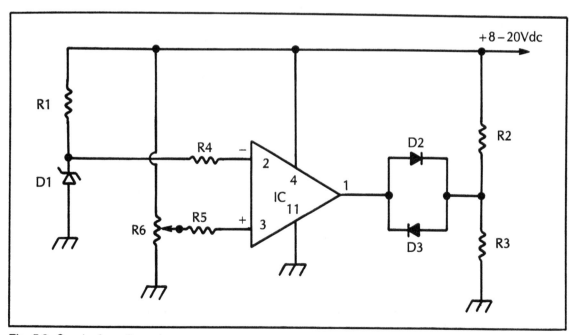

Fig. 7-2. Over/under-voltage alarm.

Table 7-2. Parts List for Fig. 7-2.

D1	5 or 6V zener
D2, D3	LED, any color
IC	LM324 quad op amp
R1, R2, R3	1 kΩ ½ W resistor
R4, R5	2.2 kΩ ¼ W resistor
R6	10 kΩ pot
Misc.	Perfboard, IC socket, etc.

output reverses its condition, lighting LED D2 to indicate an over-voltage condition. If the power source is very stable and you have a steady hand, the pot can be set to where both LEDs are turned on. Only a change of a few millivolts causes one of the LEDs to turn off.

The circuit in Fig. 7-3 offers a special feature that allows the monitored voltage to vary a preset amount without setting off the alarm. It's not uncommon for some power supplies to have their output voltage vary several percent under normal load/no-load condition. The window adjustment on this over/under circuit can be set to overlook a preset voltage variation without sending out an alarm.

The circuit works like this. The zener reference diode is isolated from loading with an emitter follower stage, Q1. The output at the emitter is connected to a 100 Ω pot and a 1 k resistor in series. The top of R9 goes to the plus input of op amp A, pin 3, and the bottom end of R9 goes to the plus input of op amp B, pin 5. The voltage across the 100 Ω pot gives an open area, or window, where neither of the op amps turns on an LED. With the pot adjusted for zero resistance, the circuit operates like the circuit in Fig. 7-2, but with the pot set to its maximum value of 100 Ω a window of approximately 1V appears before either LED lights. The input operating range is set by pot R10 in the same way as the previous circuit.

The voltage-sensing circuit can be constructed using any scheme that suits the application, but in any case the use of an IC socket makes connecting to the solid-state centipede a much safer and easier chore. The two pots and LEDs should be placed in a convenient location for ease of adjustment and visibility.

Connect the circuit across the power source that's to be monitored, and set the supply to the desired voltage. Set R9 to its minimum resistance value, and adjust R10 until both LEDs try to light at the same time. If the regulation of the supply is excellent, the window adjustment can be made so both of the LEDs remain off until an over or under voltage condition occurs. If a poorly regulated supply is monitored, the window pot (R9) may need to be set to its maximum resistance position for both of the LEDs to remain off for a normal voltage variation.

If a digital voltmeter is available, the actual window opening can be set to an exact limit, but unless an extremely accurate setting is desired a good setting can be made by jockeying back and forth between the two adjustment pots.

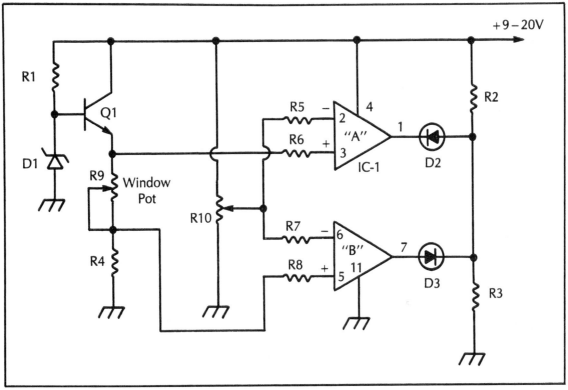

Fig. 7-3. Over/under-voltage alarm with adjustable window.

Table 7-3. Parts List for Fig. 7-3.

D1	5 or 6V zener
D2, D3	LED, any color
Q1	2N3904 npn gp transistor
IC	LM324 quad op amp
R1, R2, R3, R4	1 kΩ ½ W resistor
R5, R6, R7, R8	2.2 kΩ ¼ W resistor
R9	100 Ω pot
R10	10 kΩ pot
Misc.	Perfboard, IC socket, etc.

OVER-CURRENT ALARM

An excess current load on a power supply or battery can cause permanent damage to the supply or the equipment and place a dent in the best of pocketbooks. The circuit in Fig. 7-4 can help avoid an extended overload condition by lighting an LED as a warning.

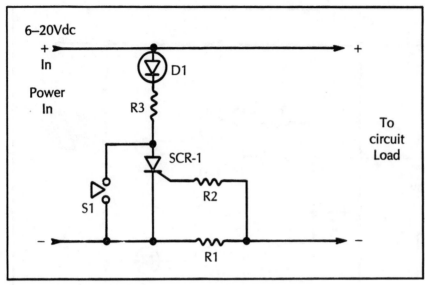

Fig. 7-4. Over-current alarm.

The current sampling circuit connects in series with the power source and load circuit. The current drawn by the load flows through the current sampling resistor R1. When the voltage across R1 reaches approximately .6V, the gate junction of the SCR draws current, turning on the device to light the LED. The value of R1 determines the current that it takes to turn on the SCR. A .6 Ω triggers the SCR when the current flow reaches about 1 A and a .06 Ω resistor causes the LED to light when about 10 A flows through the circuit.

To determine the value of R1 for any current flow, just divide .6V by the desired trip current and the result is the resistor value needed. The wattage value can be found by multiplying .6V times the trip current. The 1 A trip resistor dissipates .6 W, so a 1 W value is the nearest standard value.

Table 7-4. Parts List for Fig. 7-4.

D1	LED, any color
SCR-1	Small .6 A SCR or similar device with a 50V or greater rating
R1	R1 sets the current trip value; or a 1 A trip, R1=.6 Ω; 10 A R1=.06Ω; 100 mA trip R1=6.0Ω. etc.
R2	2.2 kΩ ½ W resistor
R3	1 kΩ ½ W resistor
S1	N.O. push-button reset switch
Misc.	Perf board, pins, etc.

OVER-CURRENT SHUTDOWN

The circuit in Fig. 7-5 goes one step farther than the previous circuit by shutting down the output of the supply until the overload problem can be addressed and corrected.

A power transistor is connected in series with the power source and the load, with a medium power transistor supplying base current to it. Resistor R3 supplies base current for the driver transistor and is also connected to the anode of a low current SCR. If the current through R1 goes too high, the SCR turns on pulling the voltage at the base of Q1 to near ground level. The emitters of Q1 and Q2 also fall to a voltage level near zero to shut down the output voltage going to the load circuit. The push-button switch resets the circuit as soon as the current drops back within its preset limit.

The current trip resistor R1 can be selected with the same simple formula as used in the previous over-current circuit of Fig. 7-4. A heat sink is needed for both transistors, and the size depends on the power supply output voltage and current drawn by the load. Any power supply can be protected by this type of current shut-down circuit as long as the power handling transistors are matched to the supply's output voltage and current loading.

The over-current alarm circuit shown in Fig. 7-6 offers a feature that allows a closer setting of the current trip value. If a load is critical as to the maximum current that it uses, this adjustable trip circuit can be set to light the LED at a small increase in the current flow into the load circuit.

The circuit operation is different than the other over-current circuits, as the LED turns on and remains on only during the time that the over-current

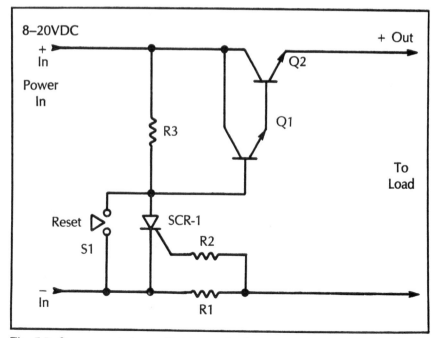

Fig. 7-5. Over-current alarm with trip-out circuit.

Table 7-5. Parts List for Fig. 7-5.

SCR-1	Small .6 A 100V SCR
Q1	2N2102 or similar medium power transistor
Q2	2N3055 npn power transistor
R1	.6 Ω fo 1 A trip, .06 Ω for 10 A, etc.
R2	2.2 kΩ ¼ W resistor
R3	100 Ω 5 W resistor
S1	Push-button N.O. reset switch
Misc.	Perf board, heat sink for both transistors, etc.

condition occurs, and goes out as soon as the current drops below the preset level. Transistor Q1's base is connected to the wiper of a 1 k pot that's tied across the current sensing resistor R1. When the voltage at the wiper increases to about .6V, Q1 begins to conduct drawing current through R4, R5, and the base emitter junction of Q2. The collector current of Q2 flows through the LED diode, lighting it to indicate a current overload condition.

The resistor value of R1 can be selected in the same manner as in the other over current circuits to cover the range needed for the actual circuit load application.

TEMPERATURE PROBE CIRCUIT

Semiconductor devices are not too forgiving if they are abused. Too much voltage, current, or high temperature can send the best device to never-never land. The last few circuits offered solutions to the first two killers, and the circuit in Fig. 7-7 can help in protecting against a smoking part. By placing the

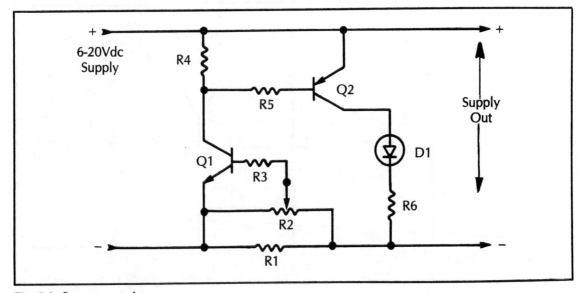

Fig. 7-6. Over-current alarm.

Table 7-6. Parts List for Fig. 7-6.

D1	LED, any color
Q1	2N3904 npn gp transistor
Q2	2N3906 pnp gp transistor
R1	Value determined by current operating range, same value as used in two previous circuits
R2	1 kΩ pot
R3, R6	1 kΩ ¼ W resistor
R4	10 kΩ ¼ W resistor
R5	3.3 kΩ ¼ W resistor
Misc.	Perfboard, pins, wire, etc.

temperature probe next to a sensitive part, the circuit can detect a rise in temperature of a few degrees and light the LED to warn that a shut-down may be necessary before a major failure occurs.

Operation of the temperature rise detection circuit goes like this. An op amp, one section of a LM324 IC, is working as a voltage comparator to weigh one adjustable input voltage against another that changes with the temperature. The circuit is set up to light an LED when the temperature of the probe goes up from a preset limit. The inverting input of the op amp, pin 2, is connected to the anode of a 1N914 silicon diode that serves as the temperature-sensitive element. The non-inverting input, pin 3, is tied to the 1 k adjustment pot that sets the temperature trigger point.

The ordinary silicon diode is very level-headed when it comes to taking the heat, and can be relied on to give out accurate information as to how hot it is. For each degree centigrade of temperature change, the forward voltage across a silicon diode changes approximately 2 mV. As the temperature increases, the voltage across the diode goes down, giving the silicon devise a negative coefficient factor.

When the voltage at the anode of D1 is slightly higher than the voltage at the wiper of R7, the op amp's output voltage is near zero, and the LED is dark. As the voltage drops ever so slightly at the anode of the diode, in relation to the preset voltage at pin 3 of the IC, the output switches to a positive voltage, lighting the LED and indicating a rise in temperature. The purpose of C1 is to keep Murphy from sending in any stray rf signals to foul up the operation. The op amp's gain is extremely high, and very sensitive to noise when both inputs are in balance, so C1 offers a negative feedback path to stop any ac signals from being amplified during this high-gain circuit condition.

Transistor Q1 and zener diode D2 form a simple voltage regulator to supply a stable voltage to the comparator's input circuitry. R1, R2, R3, and R4 are all 1 percent metal film resistors that also help maintain a stable input for the op amp. The precision resistors are not necessary if the circuit is used to monitor large temperature increases, but if a warning of a small increase is important, stick with the quality parts.

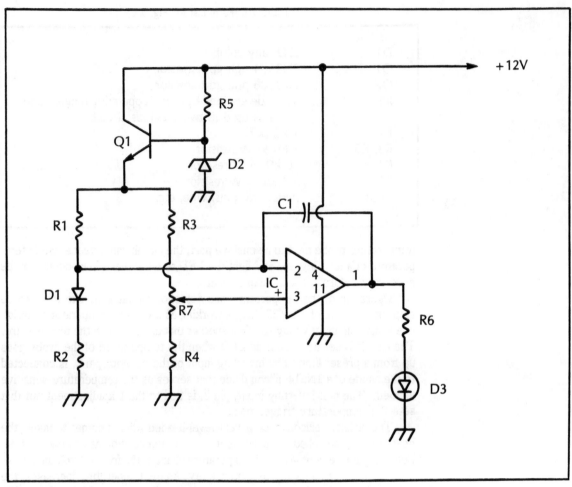

Fig. 7-7. Temperature alarm circuit.

Table 7-7. Parts List for Fig. 7-7.

D1	1N914 silicon diode
D2	6V zener diode
D3	LED, any color
IC-1	LM324 op amp
Q1	2N2222 npn transistor
C1	.047 mF 100V mylar capacitor
R1, R2, R3, R4	2 kΩ 1 percent ½ W resistor
R5, R6	1 kΩ ½ W resistor
R7	1 kΩ pot
Misc.	IC socket, perfboard, etc.

If the circuit is used to monitor the temperature of a solid-state device some distance from the comparator's circuit, the sensor diode D1 should be connected with a two-wire shielded cable. Neither lead of the diode should be allowed to make contact with the circuitry that's being monitored, but the glass envelope of the diode can be attached to the device with super glue.

Setting up the monitoring circuit is straightforward. As an example, take a power amplifier circuit that uses an expensive output power transistor that can be overdriven to a point producing a condition of thermal runaway, that will zap the device if the circuit is not shut down at once.

To obtain reliable and repeatable results with the circuit, the temperature setting pot needs to be calibrated in degrees C. Make a circular dial plate of paper large enough to note a number of temperature readings, and place it over the shaft of R7. Take the probe and insert it in a glass of water with ice cubes. Let the combination sit for a few minutes and slowly turn R7 to the point where the LED just begins to light. Mark the dial "zero C." Take the probe out of the ice water and let it warm to room temperature. Check the room's temperature with a standard thermometer and make the same adjustment with R7 and mark the dial. Place the probe in a pan of boiling water, make the adjustment, and mark 100°C on the dial.

The circuit's sensitivity can be set to pick up the warmth of the touch of your hand, or it will respond to the heat of a soldering iron several inches away. If the diode probe is mounted at the focal point in a small parabolic reflector, it can detect a heat source several feet away. With the parabolic setup, the sensor can be used to monitor a piece of equipment or a single solid-state device without actually making contact, with it.

STORM WARNING SYSTEM

The springtime brings nature's beauty in its display of flowers, green grass, leaves for the trees, and for many the dreaded severe weather that can turn a time of wonder into a nightmare. With today's modern storm warning systems the number of lives lost is down due to the accuracy and speed with which the media can get the word out. But when life and limb are at risk, any hedge that can be taken to reduce the odds of harm is worthwhile.

The storm warning alarm circuit in Fig. 7-8 may not save life or limb, but it could give a timely warning of an approaching thunderstorm. Lightning and tornadoes are the most destructive elements of any thunderstorm, and both can be detected with this alarm circuit. **Under no condition should you rely on this system or any single warning system when a severe weather condition is possible.**

How the Storm Warning Alarm Circuit Operates

Three op amps of a LM324 quad op amp are the active devices used in the alarm circuit. A 25-inch loop antenna is tuned to approximately 18 kHz or 12 kHz, according to the setting of S1, to pick up electrical disturbances caused by atmospheric radiation.

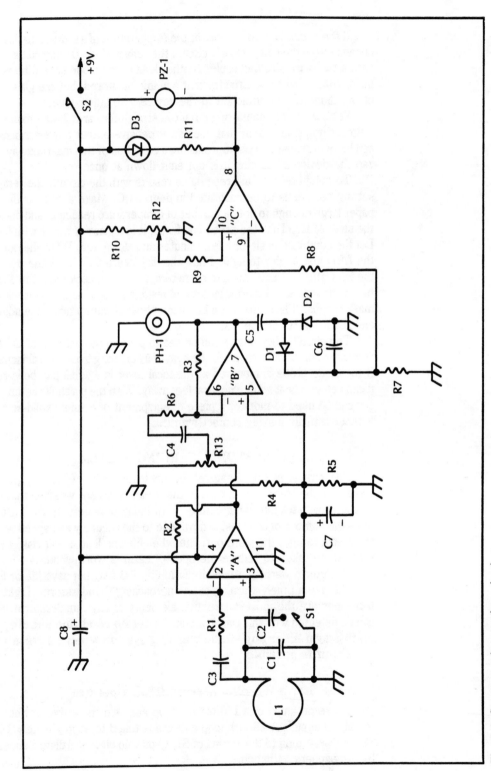

Fig. 7-8. Storm warning receiver circuit.

Table 7-8. Parts List for Fig. 7-8.

C1, C2	.005 mF 100V mylar capacitor
C3, C4, C5	.056 mF 100V mylar capacitor
C6	.1 mF 100V mylar capacitor
C7	100 mF 16V electrolytic capacitor
C8	220 mF 16V electrolytic capacitor
D1, D2	1N914 silicon signal diode
D3	LED, any color
IC-1	LM324 quad op amp IC
L1	100 turns of #26 wire on a 25-inch wood or plastic round form, one inch wide; coil can be jumble wound
PH-1	Small crystal headphone; magnetic phone will not work
PZ-1	Piezo sounder
R1	2.2 kΩ ¼ W resistor
R2, R3	220 kΩ ¼ W resistor
R4, R5	2.7 kΩ ¼ W resistor
R6, R10	4.7 kΩ ¼ W resistor
R7	100 kΩ ¼ W resistor
R8, R9	10 kΩ ¼ W resistor
R11	1 kΩ ¼ W resistor
R12	5 kΩ pot
R13	10 kΩ pot
S1, S2	SPST toggle switch
Misc.	Loop form, #26 enamel copper wire, perfboard, pins, cabinet, shielded lead, etc.

The signal picked up by the loop antenna is amplified by op amp A 100 times, and the output is connected to the circuit's gain control pot, R13. The second op amp brings the signal up by a maximum factor of 45 times, with its output connected to a high impedance piezo crystal earphone, and a voltage doubler rectifier circuit. The dc output of the doubler circuit is connected to one input of the third op amp. The other input of the op amp is connected to a threshold setting pot, R12. The output of the op amp drives a LED and a solid-state piezo sounder to give out the warning signal.

Building the Storm Warning Alarm

The best place to start construction is to wind the pick-up loop antenna. A wooden hoop 25 inches in diameter by 1 inch wide or of similar size can usually be found in a craft or hobby shop for a few dollars. Wind 100 turns of number 26 enamel copper wire around the hoop in a jumbled fashion. Tape the wire in place to keep it from coming off of the hoop, and connect the loop to the circuit with shielded cable.

The remaining component parts can be mounted on a piece of perf board with push-in pins. An IC socket for the LM324 makes the construction an easy task to complete. The circuit is non-critical, and as long as the wiring is kept neat and point-to-point, no special layout is required. The completed circuit can be housed in a small metal or plastic case with the loop separate so it can easily be rotated to determine the direction of the oncoming storm.

Unless there's an electrical storm on the way, it's hard to test the circuit's operation, but a simple check of the circuit's sensitivity and directional abilities, can be made with a TV receiver. The TV's horizontal oscillator fills the airways with a signal that the loop can pick up on and be directed toward to check out the loop's pick-up pattern.

When an electrical storm is approaching, listen in with the piezo earphone to the whistles and crackles that the storm is producing. The threshold pot should be adjusted to the point where normal noise, in a good weather period, makes the LED flicker; then back off just enough for the LED to go out and remain dark. An occasional flicker is okay, as it usually indicates man-made noise or dry weather static discharges.

MOISTURE ALARM

With the beautiful flowers that spring brings also come the rain showers that produce puddles in the yard and a mess in the basement. If that's a problem, then the circuit in Fig. 7-9 can be the solution. The circuit is set up to detect moisture and send out an alarm or turn on a water pump to lower the liquid to a level below the end of the probes. Naturally, for forty days and nights of Noah's liquid sunshine the system might not work out as planned, but for a normal spring rain the job of internal flood control can be automatically taken care of by the circuit.

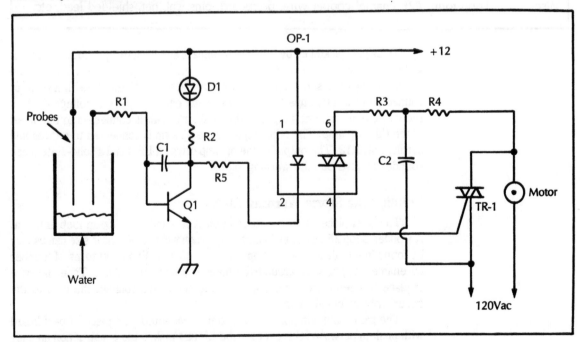

Fig. 7-9. Water level alarm.

Table 7-9. Parts List for Fig. 7-9.

C1	.01 mF 600V disc ceramic capacitor
C2	.2 mF 600V mylar capacitor
D1	LED, any color
OP-1	MOC3010 optocoupler triac driver
TR-1	6 A 400V triac, or select size to match motor size
Q1	2N3904 npn gp transistor
Probes	Metal rods of brass, stainless steel, etc.
R1	100 kΩ ¼ W resistor
R2, R5, R4	1 kΩ ¼ W resistor
R3	180 Ω ½ W resistor
Misc.	Small water pump with ac motor, perfboard, power cord, outlet socket, etc.

The circuit's operation is as simple as water falling off a duck's back. The base of a NPN transistor is connected through a current limiting resistor R1, to a metal probe. A second metal probe, parallel with the first, is tied to the positive power source. When the water level reaches the probes, a small dc current flows through the water supplying a forward bias to Q1, turning it on as a switch, and lighting the LED D1. Also, if the water pump circuit is added, the current through the internal LED of the optocoupler supplies a turn-on bias to the triac, starting the water pump.

Capacitor C1 supplies a negative feedback path to shoot down any unwanted ac signals or noise that could get into the circuit's input and cause an erratic operation with the pump motor. The current rating of the triac used should be selected to match the current requirements of the pump motor.

The two probes should be made of a metal that won't rust or tarnish to the point of losing its conductance in water. Stainless steel rods are an excellent choice to use in making the two pick-up probes, but if a less expensive route is taken, most metals work if they are cleaned from time to time. The probes should be in a parallel position, separated by about an inch, and can be kept in place with a simple spacer made of plastic or other non-conductable material.

The electronic parts are few in number and can be mounted on a small piece of perf board. An IC socket makes connecting the optocoupler IC an easy job, and a small 2×2-inch piece of aluminum will do for a heat sink for the triac. The pump's motor can be wired directly to the circuit, or an electrical socket can be used. An ac power supply with an output of 12Vdc can be used to make the circuit battery-free.

WATER DEPTH DETECTOR

How fast is the water rising? If this is a question you need answered, the circuit in Fig. 7-10 might help out. The circuit consists of three separate water level probes that can be set to any desired depth, and as the water reaches

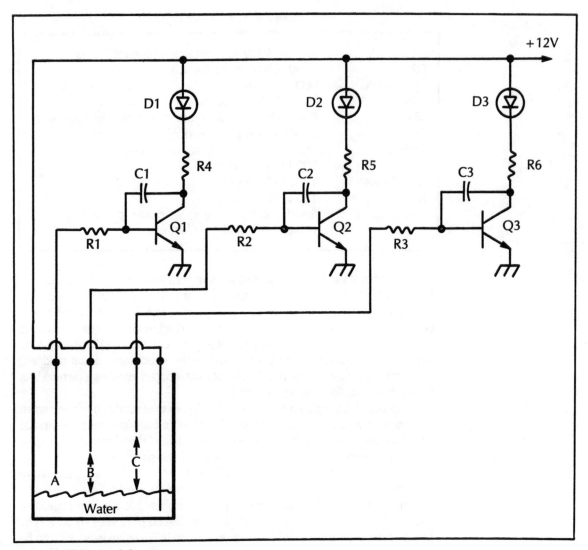

Fig. 7-10. Water level detector.

Table 7-10. Parts List for Table 7-10.

C1-C3	.01 mF 100V mylar capacitor
D1-D3	LED, any color
Q1-Q3	2N3904 npn transistor
R1-R3	100 kΩ ¼ W resistor
R4-R6	1 kΩ ¼ W resistor
Misc.	Four metal probes, perfboard, etc.

each of the probes a LED lights. A triac driver circuit like the one used in the previous circuit can be added to run a pump to keep the water at a preset level. Any number of probes with matching circuits can be added to give a more accurate reading of the water depth as it increases or decreases in level.

The circuit operates very much like the one in Fig. 7-9, without the triac driver circuit and with three LEDs, one for each of the level probes. C1-C3 offer a negative feedback path to keep the operation smooth, and the water, as it reaches each of the probes, supplies bias current to the transistor, turning on the LED in its collector's circuit.

If the circuit is to be used in a permanent location and operated continuously, an ac-operated power supply is the practical way to go.

SMOKE DETECTOR

With the great number of fire alarms available on today's market at such low prices, there's little need to add one here, but you might like to build a low cost smoke detector to put up in an area where smoking is not allowed, as a reminder to smokers. The simple circuit in Fig. 7-11 fills the bill.

An infrared diode emitter and an infrared phototransistor are used to make up the input sensor for detecting smoke or other opaque gas-like material traveling in air. Each of the infrared devices is housed in a ¼-inch × 3-inch section of opaque tubing, and positioned so one is pointing at the other, with a separation of about 1 inch between the two. The infrared light source radiated from D1 travels across the 1-inch gap between the two tubes, to the face of the phototransistor Q1, causing it to conduct and to supply a plus dc voltage at its emitter. This dc output is connected to one input of an op amp IC. The other input of the op amp is connected to a pot (R9) that is used to set the sensitivity of the circuit to respond to the desired level of smoke density that passes between the two tubes.

As long as the voltage at the emitter of Q1 is slightly higher than the voltage at the wiper of R9, the output voltage of the op amp is near zero, and the piezo sounder and LED are inactive. When smoke or an opaque material passes between the two sections of tubing, the voltage at the emitter of Q1 drops to a level below that at the wiper of R9, causing the IC to change states and activating the two warning devices.

Building the Smoke Detector

Use any construction scheme that suits, but perf board is an easy way to go, and the completed circuit, less the pick-up sensor, can be housed in a small metal or plastic case. The sounder and LED can be mounted in the cabinet's front to face toward an open area to be easily seen and heard. An IC socket should be used to avoid heat damage to the 741 op amp. The sensitivity pot R9 can be mounted to the bottom of the cabinet to allow for easy access. Drill or punch a small hole in the cabinet's back so the detector can be mounted to a wall or other solid structure.

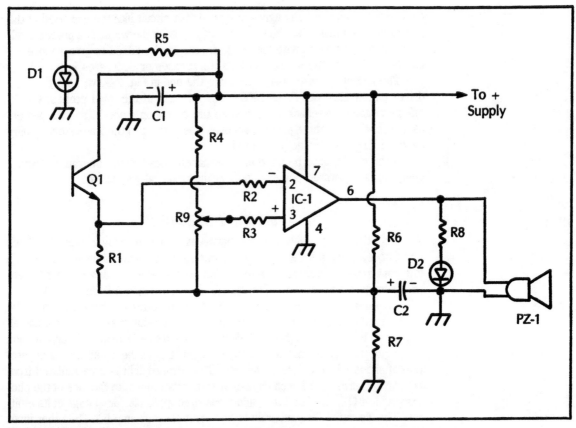

Fig. 7-11. Smoke detector.

Building the Pick-Up Sensor

A scrap piece of wood can be used to fabricate the tubing holder as shown in Fig. 7-12. The exact size is not important, but the two holes that hold the tubes in position should be drilled carefully for proper alignment. A space of 1 inch or more should be allowed between the tubing and the top section of the wooden holder so the flow of smoke won't be inhibited.

Drill two holes in the top of the wooden holder and mount the completed assembly to the bottom of the cabinet. Connect the two infrared devices to the circuitry and wrap the ends of the tubing with plastic electrical tape to keep them in place and keep room light out.

Placing the Smoke Detector in Service

Set R9 to mid-position and connect power to the circuit. If the LED and piezo are putting out an alarm signal, turn R9 to decrease the voltage at its wiper until the alarms just cease to operate. Blow smoke in the direction of the pick-up. As soon as it goes between the two sensor tubes, the alarm should go off. For a permanent location for the detector, choose a place where the normal air flow carries the smoke to the pick-up area.

142

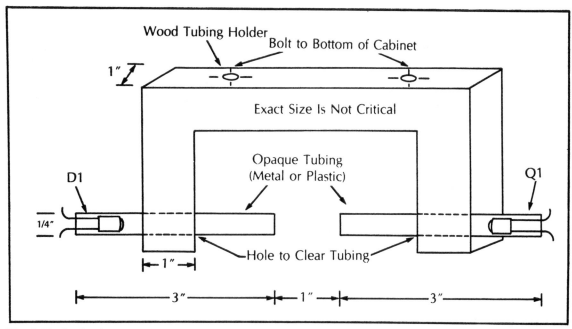

Fig. 7-12. Smoke detector pick-up drawing.

In the figure: Wood Tubing Holder, Bolt to Bottom of Cabinet, 1″, Exact Size Is Not Critical, D1, Q1, Opaque Tubing (Metal or Plastic), 1/4″, Hole to Clear Tubing, 1″, 3″, 1″, 3″

RF SNIFFER

Like it or not, electronic espionage is a way of life in the high tech world of today. A manufacturer of specialized semiconductor devices uses a number of secret and extremely valuable processes in producing a single source high dollar device. As long as the manufacturer remains the only supplier, the profits will run high, but if another firm can steal the process and compete, without

Table 7-11. Parts List for Figs. 7-11 and 7-12.

C1, C2	220 mF 25V electrolytic capacitor
D1	Infrared emitter diode, Radio Shack #276-142 or similar
D2	LED, any color
IC-1	741 op-amp IC, 8-pin mini dip
Q1	Infrared phototransistor, Radio Shack #276-145 or similar
PZ-1	Solid-state piezo electronic sounder, any frequency
R1-R3	10 kΩ ¼ W resistor
R4	16 kΩ ¼ W resistor
R5	470 Ω ½ W resistor
R6-R8	1 kΩ ½ W resistor
R9	10 kΩ pot
Misc.	Perfboard, pins, ¼-inch opaque tubing × 3 inches in length (need two), cabinet, etc.

the expense of the R and D costs, the company that developed the process could be forced out of business, or at the very least experience a large profit loss.

Even if you don't have that much to lose, but would like to know if someone is bugging your place of business or office, try the circuit in Fig. 7-13 to sniff out any rf transmitters that might be planted. The sniffer ferrets out most transmitter signals, from frequencies below the broadcast band to well over 500 mHz, and give out an audible sound when in range. The antenna can be extended for greater sensitivity, and adjusted in length for selectivity in picking up some of the higher frequency transmitters.

Here's How the Sniffer Works

The pull-up antenna picks up the rf and the 1N34A germanium diode changes it to a dc voltage that flows through L1 and R1 into the base emitter junction of Q1. Capacitor C1 passes any rf that might pass through L1 and into the base circuit of Q1 to ground. The dc signal is amplified by Q1 and Q2 sufficiently to operate the piezo sounder. A 9-volt battery supplies power to the circuit, and when no rf is being detected, the current drain is near zero.

Building the Sniffer

Pick any suitable metal or plastic cabinet that fits your needs, and cut a piece of perf board to go inside the selected cabinet. Mount the few components on the perf board, keeping all of the components' leads in the front end of the sniffer as short as possible. This includes D1, L1, C1, and R1, and the connection to the antenna. Mount the piezo sounder on the front of the cabinet along with the on/off switch. Locate the pull-up antenna along and inside one side of the cabinet and keep in place with a plastic cable clamp or plastic cable ties.

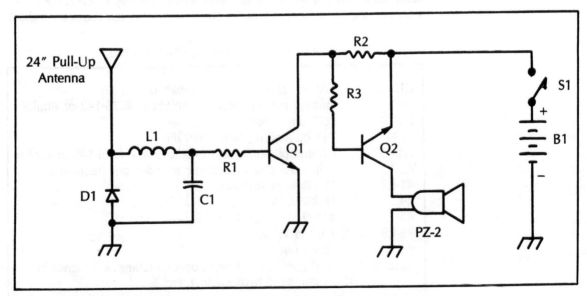

Fig. 7-13. Rf sniffer circuit.

144

Table 7-12. Parts List for Fig. 7-13.

D1	1N34A germanium diode
C1	.01 mF 100V mylar capacitor
L1	2.5 mH rf choke coil, any size close will work
PZ-1	Piezo sounder
Q1	2N3904 npn general purpose transistor
Q2	2N3906 pnp general purpose transistor
R1, R3	3.3 kΩ ¼ W resistor
R2	10 kΩ ¼ W resistor
B1	9V transistor-type battery
S1	SPST toggle switch
Misc.	24-inch pull-up antenna, small plastic cabinet, perfboard, pins, etc.

Placing the Sniffer in Service

If you have access to a wide-band rf signal generator, you can use it to check the frequency range and sensitivity of the sniffer. Loose-couple the generator's output to the antenna by laying the insulated section of the output lead across the sniffer's antenna, and sweep through the generator's frequency range, starting below 500 kHz and on up until the sniffer no longer responds. Keep the generator's output high enough to make the sniffer respond, and note the variations in the signal level needed as the frequency changes. This gives you a rough feel for its sensitivity at different rf ranges. Of course, this is not a very scientific test, but it does give you an idea as to the general operation of the sniffer.

In sweeping a room for a hidden transmitter, pull out the antenna and slowly go over the entire area. If the sniffer is to be used in an area where a powerful broadcast radio or TV transmitter is nearby, the antenna may need to be collapsed to its shortest length to be useful. Since the sniffer draws almost no current when not receiving a rf signal, it can be used as a full-time monitoring device to help keep an area clean of little clandestine transmitters.

RIPPLE DETECTOR

A regulated dc power supply can check okay when tested with a conventional voltmeter, as to voltage and current output, but can be suffering from an ailment that can raise havoc when powering a sensitive circuit, and that sickness is an ac voltage riding on top of the dc. This ripple voltage can be caused by a number of components in an early stage of failure, a drop in the input voltage, or a current overload, and in some instances can result in damage to the power supply or to the circuit it is powering if the problem isn't corrected.

Build the circuit shown in Fig. 7-14 and you can keep a constant watch on the ripple of any good power supply. The monitoring circuit connects to the output of the dc supply and couples the ac component to its input. When the

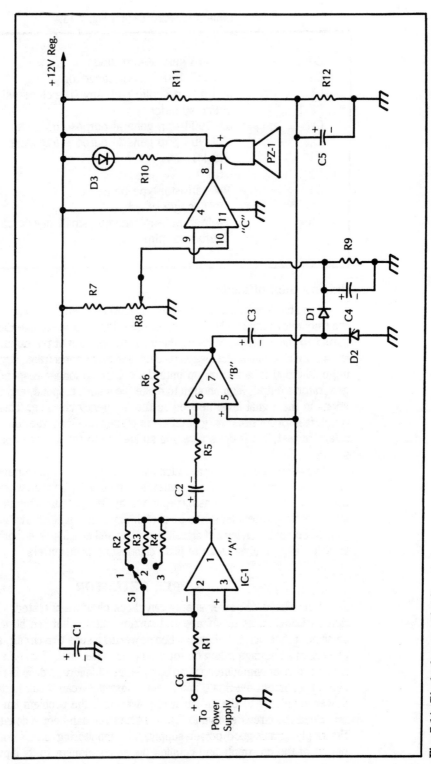

Fig. 7-14. Ripple detector circuit.

Table 7-13. Parts List for Fig. 7-14.

C1, C5	220 mF 25V electrolytic capacitor
C2	.47 mF 25V electrolytic capacitor
C3, C4	4.7 mF 25V electrolytic capacitor
C6	.47 mF 100V mylar capacitor
D1, D2	1N914 silicon signal diode
D3	LED, any color
IC-1	LM324 quad op amp IC
PZ-1	Piezo electronic sounder
R1, R4, R10, R11, R12	1 kΩ ½ W resistor
R2, R6, R9	100 kΩ ¼ W resistor
R3, R7	10 kΩ ¼ W resistor
R5	4.7 kΩ ¼ W resistor
R8	25 kΩ pot
S1	Single pole three-position switch
Misc.	Perfboard, pins, cabinet, etc.

ripple reaches a preset level, the circuit sends out an alarm to indicate the increase. The monitoring circuit can even take its power from the same source that it is keeping tabs on if the output happens to be between 9 and 16 volts, and can supply an extra few mA of current.

Ripple Detector Operation

A LM324 quad op amp performs dual duty: two of the amps add ac gain and the third operates as a dc comparator and alarm driver. Op amp A is connected in an ac gain stage to a three position switch S1, setting the feedback resistor to produce a specified gain for each position. Switch position one connects a 100 k resistor between the op amp's input and output to set the gain at 100. The second position places a 10 k resistor in the circuit, making the amp's gain 10 times. The third setting places a 1 k resistor in the feedback circuit, offering a gain of 1.

The second op amp, B, is also working as an ac amplifier with its gain set by a 100 k feedback resistor. The input resistor for this stage is 4.7 k, so 100 k divided by 4.7 k gives the stage a gain of approximately 21. With the gain switch S1 in the ×100 position, the overall gain of both stages is the gain of the first stage times the gain of the second stage, or 100 ×21, which equals a gain of 2100. If, with the circuit set for a maximum gain of 2100, a 1 mV ripple is present at the input, it is increased 2100 times to a 2.1V level which can be detected by the comparator circuit. This is the smallest ac signal that the circuit can respond to.

The amplified ac ripple signal is coupled from op amp B pin 7 to a dual diode voltage doubler rectifier circuit, D1 and D2, producing a positive output voltage, which is connected to one input of the comparator circuit, IC C pin 9. The other input of the comparator is connected to the sensitivity adjustment pot R8.

As long as the dc voltage at the reference input, pin 10, is greater in level than the output from the rectifier circuit, the output of the comparator is high and no alarm signal is given out. If the ripple voltage increases to a level that causes the dc voltage at pin 9 to become greater than the reference setting of R8, the comparator's output goes low to send out an alarm signal.

Building the Ripple Monitoring Circuit

There are two ways to go with this circuit. First, if the circuit is to be permanently connected to a specific power supply, and there's room inside the cabinet, build the monitoring circuit breadboard style with perf board and pins to fit the supply cabinet, and wire permanently in place. But if the circuit is to be used as a traveling monitoring circuit, a self-contained unit is the way to go.

No matter what construction scheme you choose to use in building the monitoring circuit, try to keep the component leads short and all of the interconnecting wiring neat and no longer in length than is necessary. As in any ac amplifier circuit, it's a good idea to keep all of the input and output leads separated from each other and away from any high level ac circuit leads. A socket should be used for the IC, and R8, D3, and the piezo sounder should be mounted in a convenient location on the unit's cabinet.

Using the Monitor

Connect the input of the monitor circuit to the power supply, with C6 going to the positive output and the circuit ground to the negative output. If the monitored power supply output voltage falls between 9 and 16 volts and there's ample reserve, the circuit can take power from the same supply.

If an oscilloscope is available, connect its input across the power supply's output and take a look to see if any ac ripple is present. Check the supply loaded and unloaded to determine the minimum and maximum levels of ripple that will be seen by the monitoring circuit.

With the circuit connected to the supply, switch S1 to position 1 and turn R9 to lower the voltage at pin 10 of the comparator until the LED and piezo sounder go off. If the circuit is operating okay and the alarm does not activate, the power supply is a dandy. With the range switch in the first position, which is the most sensitive, the circuit normally detects ripple or noise at a 1 mV level, so any supply that can pass muster with the monitor set for maximum gain should be a reliable supply indeed.

If the maximum ripple or noise figure is known for the circuit or device being powered by the supply, the monitor circuit can be calibrated to sound off when that limit is reached. If a variable output audio generator is handy, the following calibration method can be used to set the input sensitivity of the monitoring circuit.

Set the audio oscillator to generate a 120 cycle sine wave output and set the output to its minimum level. If the audio generator does not have a calibrated variable output control, use an oscilloscope or a sensitive ac amplified voltmeter to read the generator's output.

S1 switch position number 1 handles ripple signals up to about 3 mV; position number 2 signals up to about 30 mV; and position 3, approximately 300 mV. If a larger ripple signal is to be monitored, the value of R1 or R5 can be increased to lower the overall gain of the circuit. By increasing R5 to a value of 100 k, the gain of the second amplifier drops to one. This allows the input signals to be twenty times greater in level than with the original gain setting of twenty. This lets the first switch position monitor a ripple signal up to 60 mV, position 2 up to 600 mV, and position 3, a signal as great as 6V.

Set the gain switch to the position that includes the ripple level allowed for the supply used, and adjust R8 for a maximum voltage at its wiper. Connect the circuit's input to the output of the audio generator, and set the generator's output for the desired maximum ripple level. Turn R8 to lower the voltage at its wiper until the alarm sounds off. This concludes the simple calibration of the monitor circuit.

TELEPHONE IN-USE ALARM

If you have a number of extension phones on a single line and with several other people sharing the phone time, it's hard to tell, without listening in, if the line is busy or not. Not only is it bothersome to keep guessing when to grab the phone to see if it's free or still busy, but it interferes with the parties using the phone. The telephone in-use alarm circuit of Fig. 7-15 can help solve the phone guessing game. Although it won't give anyone more time on the phone, it can make using the telephone a little more convenient.

How the Telephone In-Use Alarm Circuit Operates

A two-transistor circuit senses the voltage across the telephone lines. If the phone line is inactive, the voltage across the line, tip to ring, is about 48Vdc. When the line is in use, the voltage drops to less than 10V. The alarm

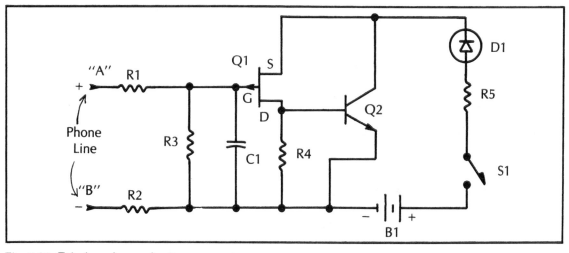

Fig. 7-15. Telephone in-use circuit.

Table 7-14. Parts List for Fig. 7-15.

B1	9V battery
C1	.02 mF 100V mylar capacitor
D1	LED, any color
Q1	2N4360 FET transistor
Q2	2N3904 npn general purpose transistor
R1, R2	10 MΩ ¼ W resistor
R3	22 MΩ ¼ W resistor
R4	39 kΩ ¼ W resistor
R5	1 kΩ ¼ W resistor
S1	SPST toggle switch
Misc.	Perf board, pins, wire, battery snap, cabinet, etc.

circuit is isolated from the phone line with two 10 MΩ resistors in the input circuit, so no interference occurs with the telephone system. The other ends of the two resistors connect to the gate circuit of a 2N4360 FET transistor. A 22 MΩ resistor completed the dc current path, and a .02 μF capacitor removes any noise that might get into the FET's gate circuit.

As long as the 48V is present on the phone line, the FET's gate is biased off with the positive voltage, but when a phone is in use the line voltage drops to less than 10V and the FET turns on, supplying base current to the 2N3904 transistor and lighting the LED. The LED stays on as long as any of the phones on the same line are in use.

Building the Telephone Alarm Circuit

The circuit is so simple you can select any construction scheme that suits, but the standby perf board and pins make a quick job of it. The circuit can be housed in a small plastic cabinet with the switch and LED located in a handy place on the top or front of the enclosure. The leads connecting to the phone lines can be wired into one of the new mini phone plugs to plug in to a standard phone outlet. To make the alarm system 100 percent effective, an alarm circuit should be used at each of the phone locations so no one needs to bother anyone else when using a phone.

If you don't want to use batteries as the power source for the monitor, a wall plug-in power supply can be used, but be sure that its output is well-filtered, as many only supply a raw dc output.

Using the Telephone Alarm

The first thing to do is to determine the polarity of the phone lines, and make a note of which lead is positive and which is negative. Then connect the positive phone line to the A input and the negative phone line to the B input of the alarm circuit. Use S1 to turn on the power, and with the phone circuit not in use the LED should be dark. Lift the handset off-hook and the LED should light up. If so, you are ready to install a similar alarm circuit at each phone location.

Chapter 8
Electric Motor and Internal- Combustion Engine Alarm Circuits

The majority of the manual labor that man performed 100 years ago is now handled by electric motors and, on a lesser scale, internal-combustion engines. Even the horse that took a load off of man's back is now, in most cases, used for pleasure with the spare time given to man because of the modern power replacements. Take a look at any modern manufacturing facility, and the number of electric motors powering equipment runs into the hundreds. They range in size from the fractional horsepower motors that drive hand operated drills, screwdrivers, etc., to the one hundred horsepower giants that run the punch presses, and other power demanding machinery.

The monitoring and protection of electric motors and internal combustion engines is another area where electronics can do an excellent job. Although your electric washing machine or even your favorite riding lawn mower are not really, in most cases, valuable enough to add a monitoring circuit, where an electric motor or engine is used to perform a critical job, or where a failure would be very costly, a monitoring circuit can pay off in spades.

AC MOTOR CURRENT SENSOR

The following example shows just how valuable an electronic monitoring circuit can be for an electric motor. A local manufacturing company uses a special process to produce a product that requires a high-volume exhaust fan to keep the flammable fumes from collecting in the exhaust tower. On one hot August afternoon, the temperature of the fan motor rose too high, causing it to slow down over a period of time and allowing the fumes to accumulate in the tower. No one is positive what caused the ignition of the fumes, but everyone present that day can certainly give an account of what happened. Fortunately for all no one was hurt, and the fire was under control in only minutes thanks to the local fire department which only had to travel two blocks to reach the factory.

Not all similar failures end with as little loss as this one, and it's true the production line was closed for a few days for repairs, but under other conditions the complete factory could have been destroyed. Within one week after the fire a special monitoring circuit was added to the exhaust fan motor to give out an alarm so the production line could be shut down before the exhaust fan completely failed.

If a clamp-on ac current meter had been connected on one of the power leads going to the fan's motor the day of the fire, it would have indicated a steady increase in the operating current of the motor. This clue would have indicated trouble that could have sent a message to an electronic sensor and alarm circuit to warn of the failure and avoid the equipment and time loss that occurred.

The block diagram in Fig. 8-1 illustrates just how that can be accomplished. An ac current sensing circuit is connected between the motor and the ac line. A sensitivity adjustment pot allows the setting of the desired current limit value that's required to set off the alarm circuit.

The heart of the monitor circuit is the device that measures the ac current flow through the motor. There are at least two methods in which the current can be measured. One method is to use a high wattage power resistor in series with the motor circuit and measure the voltage drop across it for the current flow. The problem with this method is the power loss and heat generated by the resistor. Another, and more practical, method of measuring the current flow is with an ac current transformer. A real plus for using this method is that a current transformer can easily be fabricated in the home workshop for less than ten dollars, and even less if your parts supply includes the right stuff.

A basic current measuring circuit is shown in Fig. 8-2, where the current transformer is made out of an old low-voltage transformer. T1 started out in life as a 24-volt 2-amp general purpose power transformer available at most radio supply houses for about six dollars. The transformer's secondary winding must be the outside winding so it can be removed without taking the transformer apart. The simplest method to use in removing the winding is to unsolder the outside lead of the secondary and unwind a single turn at a time. This method is slow, but it helps keep any damage from occurring to the primary winding.

With a motor connected as shown in Fig. 8-2 in series with the current transformer and power line, a relative indication of the current flow through the motor can be read on the ac voltmeter monitoring the voltage across the 120V winding. The meter can be calibrated for a normal and a danger operating area, but the real problem with this scheme is that a live body must be at hand to keep an eye pealed on the meter continuously to avoid a serious problem.

The circuit in Fig. 8-3 keeps a silent vigilance on the motor's operation and cries wolf only when a problem starts to build. That still means someone must

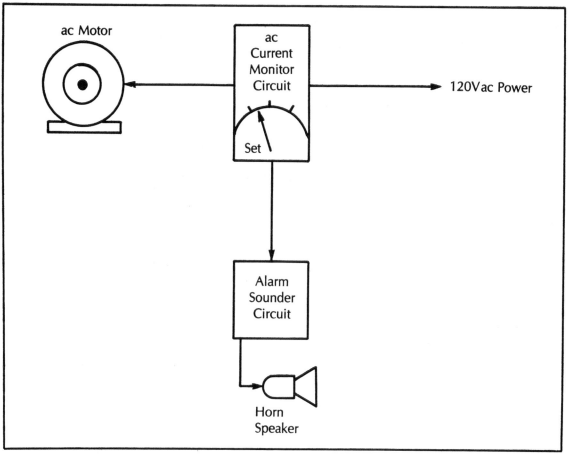

Fig. 8-1. Block diagram of ac motor current monitoring sensor.

be in hearing range of the alarm sounder, but they are not tied down to that job alone.

How the Ac Motor Monitoring and Alarm Circuit Operates

The turns ratio of the current transformer is several hundred to one, from the new two turn winding to the original 12V primary winding. The motor current flowing through the two turns winding only produces a few mV drop across it, but with the high step-up ratio of the transformer, the voltage on the 120V side is 1 to over 10V, depending on the actual current flow through the motor.

Four 1N4002 diodes are connected in a full wave bridge rectifier circuit that converts the ac voltage to dc and the filter capacitor C1, takes the humps out. The dc signal voltage connects to the input of a voltage follower op amp circuit that adds isolation and allows the use of a start-up delay circuit. The starting current of most ac motors is usually much greater than the normal running current. If the monitoring circuit looked at the starting current each time the power is applied, the alarm would go off until the motor reached its operating speed.

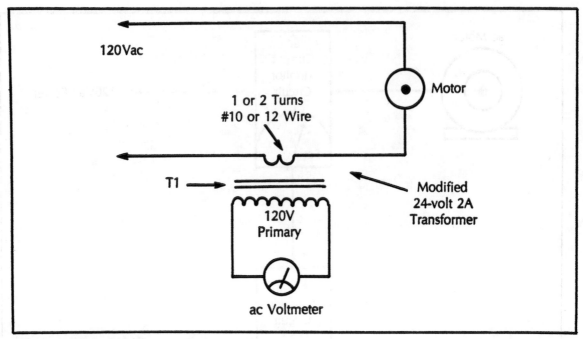

Fig. 8-2. Simple motor current monitoring circuit.

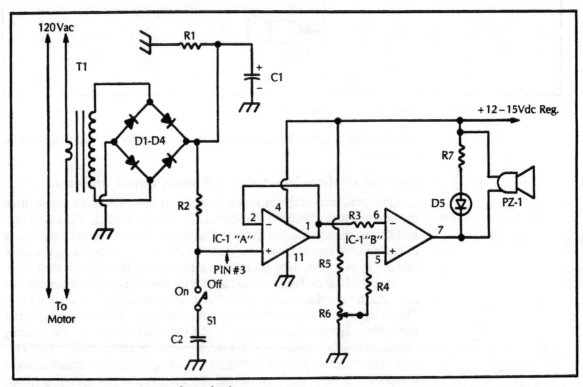

Fig. 8-3. Ac motor over current alarm circuit.

Table 8-1. Parts List for Fig. 8-3.

C1	220 mF 50V electrolytic capacitor
C2	.5 mF 100V mylar capacitor
D1-D4	1N4002 1 A silicon diode
D5	LED, any color
IC-1	LM324 quad op amp
R1, R7	1 kΩ ½ W resistor
R2	2.2 MΩ ½ W resistor
R3, R4	2.2 kΩ ¼ W resistor
R5	2.7 kΩ ¼ W resistor
R6	10 kΩ pot
PZ-1	Piezo sounder
S1	SPST switch
T1	Current transformer, made from 24V 2 A transformer (see text)
Misc.	Perfboard, pins, IC socket, cabinet, etc.

S1 gives the option of a start-up delay or an instant-on action to the circuit. With S1 in the off position, the alarm responds almost instantly if the starting current is over the set operating limit. Placing S1 in the on position gives the motor several seconds to come up to speed and settle down to a normal operating current before the circuit can send out an alarm. The value of C2 can be increased to add time to the delay circuit if needed, and if a shorter time is desired, lower the value of R2.

The output of the voltage follower pin 1 connects to one input of a voltage comparator circuit at pin 6 of IC1-B. The comparator's other input is tied to the wiper of a variable resistor (R6) that supplies a dc reference voltage for setting the current limit for the alarm circuit. Under normal operating conditions, the voltage from the current transformer, after it is converted to dc is less than the dc voltage at the comparator's reference input, and the voltage at the output, pin 7, is near that of the power supply. The voltage across the LED and piezo sounder is zero and no output is given, but when the dc voltage from the current transformer is greater than the limit setting, the comparator's output switches to ground level and the full supply voltage is applied to the LED and piezo sounder circuit. An alarm is given out.

Just about any good stable power source with an output of 12- to 15V at 50 mA can be used to operate the circuit, or, if a stand-alone unit is desired, a simple regulated supply can be built and placed in the same cabinet.

Building the Motor Monitoring Circuit

The complete circuit can be built on a small piece of perf board and housed in a metal or plastic cabinet. To reduce the chance of heat damage to the LM324, use an IC socket. A heavy duty power cord and outlet for the 120Vac input and output circuits should be included, with the outlet mounted in a convenient location on the cabinet. Mount S1, R6, D5, and the piezo sounder away from the 120Vac power circuitry, but in an area where each is easy to see and adjust.

Fabricating the current transformer (Fig. 8-2) is an easy job that can be completed in about an hour's time. A 24V 2A transformer, with the secondary winding wound on the outside of the 120V primary, was modified for the current transformer used for T1, but any similar transformer with a 120V primary and a low voltage secondary that is wound as the outer winding can be used. The main consideration is to have ample room, after the low voltage secondary has been removed, for the new two-turn winding to fit in its place. Use a piece of number 10 or 12 solid copper wire, with good insulation, for the two-turn winding.

Using the Ac Motor Monitoring Circuit

Plug the motor into the outlet on the monitor and connect dc power to the circuit. Place S1 in the off position, and set R6's wiper to ground potential. Plug the ac input to a 120V line and as the motor starts and runs, the LED and piezo sounder responds with an output. Turn R6 in the direction that increases the voltage going to the comparator, until the LED and piezo sounder just cease operation. With the sensitivity set at this point, any little change can cause the alarm to go off, and unless the motor is performing a very sensitive function the sensitivity control can be set for a greater input current change.

If the motor is turning a fan or performing any other function where an external load can be applied, the alarm trip sensitivity can be set to respond only to a real overload condition. While the motor is operating with a normal load, adjust R6 for a maximum output voltage, and, if a fan is the normal load, carefully restrict the air flow reaching the fan to increase the motor's operating current. Here's where a clamp-on ammeter would be handy to check the normal and the overload current values.

Keep the external load in place and slowly turn R6 to lower the voltage at its wiper until the LED and piezo sounder turn on. Remove the load and the alarm should go off. Stop the motor, and place S1 in the delay (or on) position. Start the motor, and if the time delay is long enough, the alarm will not activate. If the alarm does activate for a few seconds, the value of C2 can be doubled to extend the start-up time. Never use an electrolytic capacitor for C2, or to add to the value of C2, because the leakage current might be too great and the timer would never reach a timed-out condition.

If more than one ac motor requires monitoring in the same general area, then two circuits can easily be constructed on the same perf board. Use the remaining two op amps, C and D, for the second alarm circuit. The cost of adding the second circuit is several dollars less than building a complete unit for the second motor.

OVER-RPM INTERNAL COMBUSTION ENGINE ALARM CIRCUIT

One of the most damaging things that can happen to an internal-combustion engine is to have the governor fail, allowing the engine's speed to build rapidly into a runaway condition that usually ends in the destruction of the engine, and damage to surrounding items. Even Dad's old lawn mower has a built-in governor, coupled with the throttle, that keeps the engine from going over the maximum RPM's and committing hara-kiri.

Most small engines are not valuable enough, or performing a critical enough service, to bother with adding an electronic circuit to warn of an over-RPM condition, but there are conditions and special equipment operated by internal-combustion engines that can use the monitoring alarm system. One example where the alarm can be useful, is to monitor a standby 120Vac gasoline engine-driven generator that supplies ac power in an emergency. Not only can the engine sustain damage from an over RPM operation, but equipment powered by the generator can also be damaged.

If a gas engine is used to drive a special piece of equipment that must not exceed a specified RPM, then the monitoring circuit can be used to notify the operator that an over-RPM condition is starting to occur, and corrections need to be made before any damage can occur. Anything that turns, no matter what the power source might be, can use an over-RPM monitoring circuit to give out a warning in time to make adjustments, or to make it through the nearest exit to a more safe location.

The circuit in Fig. 8-4 can be set up to tattle on anything that turns when it speeds up past the set alarm limit. An infrared LED supplies a source of light that's reflected off a piece of white tape or paint located on the turning object and is picked up with an infrared phototransistor.

Here's How the Over-RPM Alarm Circuit Operates

The IR light reflected off of the rotating object appears as a short light pulse to the IR photo-sensitive transistor. The output at the collector of the phototransistor is a slightly rugged negative pulse. Transistors Q2 and Q3 shape and amplify the pulse to a uniform shape and proper polarity that's suitable for triggering the monostable multivibrator IC-1. The multivibrator's positive pulse output time period is set by the values of R7 and C1, and for every input pulse the multivibrator produces a precise timed output pulse at pin 3 of IC-1. The pulses coming out of the multivibrator are fed to a R/C voltage averaging circuit, R8 and C4.

The averaged pulse voltage is fed to a voltage follower circuit, IC-2A, that isolates it from any loading. The output, pin 1 of IC-2A, feeds one input of a comparator circuit, pin 6 of IC-2B. A variable reference voltage is taken off at the wiper of R13 and is fed to the other input of the comparator, pin 5. The comparator's output drives the alarm sounder and LED to indicate an over-RPM condition.

Power for the circuit can come from any 10- to 12Vdc source that can supply about 50 mA of current, and the output voltage should not vary more than 5 percent.

Building the RPM Alarm Circuit

The circuit is straightforward and non-critical, so any good construction scheme can be used. The pick-up can be built in the same cabinet as the re-

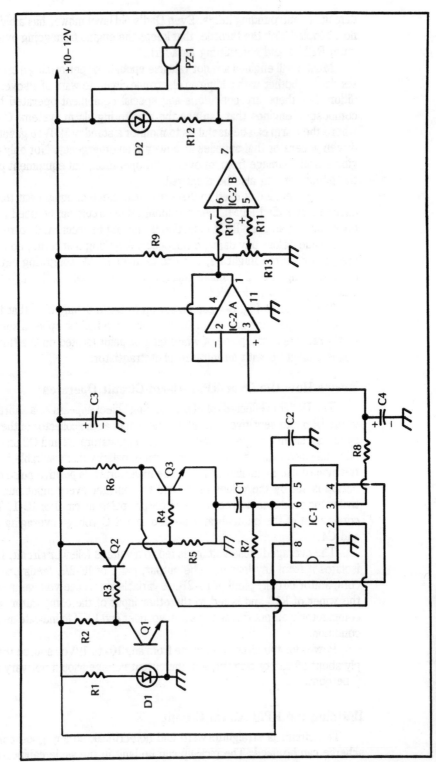

Fig. 8-4. Over-RPM alarm circuit.

Table 8-2. Parts List for Fig. 8-4.

C1	.056 mF 100V mylar capacitor
C2	.1 mF 100V mylar capacitor
C3	470 mF 25V electrolytic capacitor
C4	6.8 mF 35V tan capacitor
D1	Infrared emitting diode, Radio Shack #276-142
D2	LED, any color
Q1	Infrared phototransistor, Radio Shack #276-145
Q2	2N3906 pnp general purpose transistor
Q3	2N3904 npn general purpose transistor
IC-1	555 timer IC
IC-2	LM324 quad op amp IC
PZ-1	Piezo sounder
R1	470 Ω ½ W resistor
R2	10 kΩ ¼ W resistor
R3, R4, R10, R11	2.2 kΩ ¼ W resistor
R5	4.7 kΩ ¼ W resistor
R6, R12	1 kΩ ½ W resistor
R7, R8	100 kΩ ¼ W resistor
R9	2.7 kΩ ¼ W resistor
R13	10 kΩ pot
Misc.	IC sockets, perfboard, pins, cabinet, etc.

maining circuitry, or can be separate and in a smaller cabinet of its own. In either case, the IR devices should be shielded from other light sources with an opaque enclosure (see Fig. 8-5). Two short pieces of a soda straw painted black could be used to house the two IR semiconductors in, or a block of plastic or wood could be drilled to let each device slip in place. In any case, the two devices should be placed at an angle to focus on an object no farther away than one inch, and closer if possible. By making the focal distance as close as possible there is little chance of an outside light source causing any interference with the IR reflected signal.

Sockets should be used for both ICs, and the parts can be mounted on a section of perf board push-in pins, or a circuit board can be made if several circuits are to be produced. The RPM limit-setting pot R13, D2, and the piezo sounder should mount in a convenient location on the cabinet's top or front section. If the pick-up is to be used several feet away from the main circuitry, then shielded leads should be used for both IR devices.

Using the RPM Alarm Circuit

A small section of white tape ½-inch square, or a dab of white paint covering the same area, can be attached to the revolving object. If the turning object

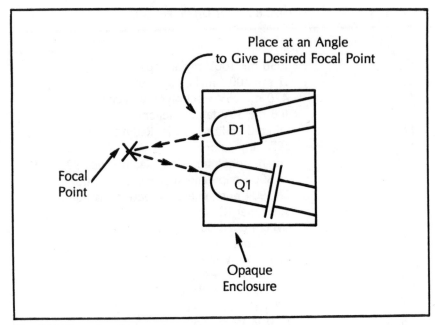

Fig. 8-5. IR semiconductor setup.

happens to be too reflective, there might not be enough contrast between it and the added white material to give a good output pulse. If this is the case, then the revolving object needs to be coated with a no gloss non-reflective coating of paint or similar material. If the object cannot be coated; a circle of dark paper can be cut to fit and then attached to the item, with a white stripe added to the paper to reflect the IR light source. In most installations, some experimenting is necessary to obtain the best operating results with the circuit.

If the speed of the revolving object requires the over-limit setting of R13 to be at either extreme end of its rotation, then the values of R7 and C1 should be changed to bring the normal operation range back toward the mid-range adjustment of the pot. When the limit setting falls too close to the low voltage end of the adjustment pot, the value of C1 or R7 can be increased to bring the level up, and if the adjustment is near the upper voltage level, lower the values of C1 or C7 to move the setting down.

Connect the positive lead of a dc voltmeter, set on the 10V scale, to pin 1 of IC-2A while the engine is running and while moving the pick-up back and forth to determine the inner and outer operating limits. The voltage drops in value as the pick-up is moved beyond its useful limits in either direction, and the mid position between the two limits is an ideal location for the pick-up.

With R13 set to produce a maximum voltage at its wiper and with the engine running at the desired speed, slowly turn R13 to lower the wiper's voltage until the alarm goes off. Turn the pot back about 1/16th turn past the point where the alarm ceases for the new limit setting. The alarm goes off if the speed increases past the limit setting. If a portable tachometer is available, use it to read the engine's normal speed, and to set the new limit speed to a known RPM. Either method of setting the alarm's limit point detects an engine that is beginning to speed up and possibly go into a runaway condition. The early detection and correction of an engine ready to self-destruct is the bottom line, and the over-RPM alarm circuit can help.

Chapter 9
Selecting, Building, and Installing an Alarm System

In selecting an alarm system, the number one requirement is to have something of value to protect that is worth the expense and time necessary to properly plan, build, and install the system. For the majority of us, our homes and automobiles are the most valuable material items that we own and owe for. These two items can be protected against fire and theft in a number of ways, but neither can be completely protected against a professional thief. You must remember that the successful professional thief works very hard and long at his or her trade, with more incentive to succeed than most other professions, because success means above everything else, not getting caught in the act or later by evidence left because of a sloppy job.

The only chance you have to outfox the professional thief is to install an alarm system that he's never seen the likes of before. It doesn't take many surprises before the professional thief quickly moves on to easier pickings. Nothing looks better to the professional thief than the old standard burglar alarm installation. This is duck soup for the pro, who can cut right through the system and make off with your treasures without a hitch. Just remember, to catch a fox, set a unorthodox trap.

INSTALLING A BURGLAR ALARM

Spend ample time planning your system before spending a dime on special equipment or getting your tools out for the installation. The most sophisticated high dollar alarm system you can buy is of no value if it is improperly installed.

Time spent planning, even on the simplest system, is many times more valuable in ending up with a good system than the many dollars spent for special alarm equipment without a clue of where and how the installation should be done. Always plan first, and then put together the system's component parts, and the final step of making the installation will go without a hitch.

After you have planned your complete alarm system and have the component parts gathered together, it's time to start the installation of the parts to combine into a complete working alarm system that will not only fool the amateur, but the professional thief as well.

Probably the best all-around alarm system to install to trap both types of thieves is a dual sensor system. The obviously-located sensors let the rank amateur know that the property is protected and he stands a good chance that the job will end in a misadventure. This alone can keep many would-be burglars from breaking in and damaging property. It also lets the pro know that there is some type of protection onboard.

The second line of defense is the well-planned and installed sensor that even the pro cannot easily detect. All it takes is for one of the sensors to pick up on the intruder and set off the alarm to make the installation job worth while. The well-hidden sensor is no better than a poor one if the wire leads leaving it show up to the pro like a red flag to an angry bull. In this game, it's the small details that separate the amateur from the professional on both sides of the law.

If variety is the spice of life, it can also be the speck of pepper in a thief's eye. If the majority of sensors used in a single alarm system are different, then the chance of anyone getting by all of the sensors is almost nil. Keep 'em guessing!

CREATING YOUR OWN ALARM SYSTEM

That's what this book is all about. If you build the total alarm system, only you will really know the ins and outs of the complete system, so right off you are one up on the would-be thief. Always remember that you are but one person with limited means at your disposal to protect your worldly goods, and the predators of the world are many, so keeping the edge on the competition is the name of the game in the good guys vs. the bad guys business.

A number of useful sensor and alarm circuits in Chapter 2 deal with the second most valuable item that many of us own today, and that's the family automobile. Having your automobile stolen is not the only costly personal catastrophe that can happen. A blown engine caused by a faulty idiot light can end up costing you much more than it would cost if the car was stolen and replaced in full by your insurance company.

Any protective device that can be added to an automobile that gives out an early warning signal that something is about to fail can be like an insurance policy that money can't buy. So, if that heap you are driving and the wallet you're sitting on means a lot to you, then build your own electronic sensors and alarm system to give you the leading edge over the manufacturer's minuscule warning system.

Look over each of the preceding chapters closely and hopefully you will find a number of circuits that will be useful in helping you to solve a few of life's problems electronically.

INDEX

INDEX

Other Bestsellers of Related Interest

**TROUBLESHOOTING AND REPAIRING
SOLID-STATE TVs—2nd Edition**
—Homer L. Davidson

With this updated, complete workbench reference, you will have practical information on troubleshooting and repairing all the most recent solid-state TV circuitry used by the major manufacturers of all brands and models of TVs. This new edition includes the latest material on high-definition TV, or HDTV, and spike bar protectors, as well as in-depth looks at particular circuits from Sylvania, RCA, Radio Shack, and Panasonic televisions. 624 pages, 698 illustrations. Book No. 3700, $24.95 paperback, $36.95 hardcover

**TROUBLESHOOTING AND REPAIRING VCRs
—2nd Edition**—Gordon McComb

This book has helped more than 80,000 VCR owners keep their machines working at peak performance. With this book and a basic set of tools, you can handle most VCR problems quickly and easily—from simple parts cleaning and lubrication to power supply repair and circuitry malfunctions. This revised second edition updates the bestselling original volume with the most recent technological advances. 432 pages, 186 illustrations. Book No. 3777, $19.95 paperback, $32.95 hardcover

**TROUBLESHOOTING AND REPAIRING
CAMCORDERS**—Homer L. Davidson

This superb troubleshooting guide shows you how to repair any brand of VHS, VHS-C, Beta, or 8-millimeter video camera on the market today. Davidson provides clear instructions along with diagrams and service literature from a wide variety of manufacturers, plus hundreds of schematics to speed diagnostics and repair. Some of the many topics covered include cleaning and lubricating camcorders, system control circuits, audio circuits and microphones, and the various motors in a camcorder. 544 pages, 606 illustrations. Book No. 3337, $22.95 paperback, $35.95 hardcover

**TROUBLESHOOTING AND REPAIRING
AUDIO EQUIPMENT**—Homer L. Davidson

When your telephone answering machine quits . . . when your cassette player grinds to a stop . . . when your TV remote control loses control . . . or when your compact disc player goes berserk . . . you don't need a degree in electronics or even any experience. Everything you need to troubleshoot and repair most common problems in almost any consumer audio equipment is here in a servicing guide that's guaranteed to save you time and money! 336 pages, 354 illustrations. Book No. 2867, $18.95 paperback only

**20 INNOVATIVE ELECTRONICS PROJECTS FOR
YOUR HOME**—Joseph O'Connell

More than just a collection of 20 projects, this book provides helpful hints and sound advice for the experimenter and home hobbyist. Particular emphasis is placed on unique yet truly useful devices that are justifiably time- and cost-efficient. Projects include a protected outlet box (for your computer system), a variable AC power controller, a remote volume control, a fluorescent bike light, and a pair of active minispeakers with built-in amplifiers. 256 pages, 130 illustrations. Book No. 2947, $13.95 paperback only

**ROBOT BUILDER'S BONANZA: 99 Inexpensive
Robotics Projects**—Gordon McComb

Where others might only see useless surplus parts you can imagine a new "life form." Now, there's a book that will help you make your ideas and dreams a reality. With the help of *Robot Builder's Bonanza* you can truly express your creativity. This fascinating guide offers you a complete, unique collection of tested and proven product modules that you can mix and match to create an almost endless variety of highly intelligent and workable robot creatures. 336 pages, 283 illustrations. Book No. 2800, $16.95 paperback only